LADYBIRD BEETLES as "Bio Security Force"

The Authors

Ponnusamy N. graduated from the Horticultural College and Research Institute, Tamil Nadu Agricultural University, Periyakulam with first class. He obtained his M.Sc. (Ag) in Uttar Banga Krishi Vishwavidyalaya, Cooch Behar, West Bengal. For his Post graduate studies, he got National Talent Scholarship (NTS) from ICAR. At presently, he is pursuing his Ph.D in Dr. Rajendra Prasad Central Agricultural University, Pusa, Bihar. He attended several international and national conferences as well as workshops.

Nagendra Kumar is working as Assistant Professor-cum-Scientist in AICRP on MAP & Betel vine, Department of Entomology, Dr. Rajendra Prasad Central Agricultural University, Pusa, Bihar since 2010. Now, he is perusing Ph.D (In-service) at UBKV, Pundibari, Cooch Behar (W.B). He has qualified ICAR NET. He obtained M.Sc.(Ag.) and B.Sc.(Ag.) in Entomology & Agril. Zoology from BHU, Varanasi in 2000 and 2002, respectively. He actively involved in research, teaching and extension. He published 14 research papers in National and International Journals, 6 book chapters and more than 15 popular article. He honored with best teacher award in 2017 and Young Scientist award in 2017 & 18. He is also editorial board member of *International Journal of Applied Agricultural Sciences and International Journal of Zoology and Animal Biology* (IZAB).

Anil Kumar is presently working as Assistant Professor-cum-Scientist at SRI, Department of Entomology, Dr. Rajendra Prasad Central Agricultural University, Pusa, Bihar. He did his B.Sc.(Ag.) and M.Sc.(Ag.) in Entomology and Agril. Zoology at BHU, Varanasi (U.P.) with first class during 2002 and 2004, respectively. Now, he is perusing Ph.D at UBKV, Pundibari, Cooch Behar (W.B). He has qualified ICAR NET and SRF. He joined as Assistant Professor-cum-Scientist (Entomology) at July 2007. He has more than 30 research article and more than 15 popular articles to his credit and also has many book chapters in different aspect. He has got best teacher award 2017 by Society for Upliftment of Rural Economy and young scientist award 2018 by Agro - Environmental Development Society.

LADYBIRD BEETLES as "Bio Security Force"

– Authors –

Ponnusamy N.

Nagendra Kumar

Anil Kumar

2020

Daya Publishing House®

A Division of

Astral International Pvt. Ltd.

New Delhi – 110 002

ISBN 9789390371556 (Int. Edition)

Published by : **Daya Publishing House®**
A Division of
Astral International Pvt. Ltd.
– ISO 9001:2015 Certified Company –
4736/23, Ansari Road, Darya Ganj,
New Delhi-110 002
Ph. 011-43549197, 23278134
E-mail: info@astralint.com
Website: www.astralint.com

Printed at : **Raplika Press Pvt Ltd**

Preface

Crop production around the world is under constant threat by various pests. There are different pest management tactics, in which biological control is eco-friendly and safe for non target organisms. Coccinellids are plays important role in biological control. It's act as a friend to farmers due to their feeding efficiency of crop pests.

In this concern, we have written the book "***Ladybird beetles as Bio Security Force***". This book deals with the significance of predatory coccinellids in pests control. It's consists morphology, life cycle, taxonomy and distribution, mating and reproduction, communication, prey specificity, predatory guilds, conservation techniques, mass culturing and natural enemies of coccinellids.

We hope, this book will be useful for students, teachers and farmers. We thank the "Astral International (P) Ltd, New Delhi" for bringing out the book.

Ponnusamy N.
Nagendra Kumar
Anil Kumar

All for One and One for All, Keep Biodiversity or Our FUTURE may Fall

Dedicated to Nature

Contents

Chapter 1

Introduction

The arthropod insect pests attacked on different crops, which fade the quality of crops and reduce yield as well in the economy of farmers. Farmers used agrochemicals for the protection of crops by the attack of insect pests, which increase crop production. Pesticides create many serious problems such as air, water, and food pollution; health hazards, ill effects and killing of beneficial organisms, pest resistance, pest resurgence, secondary outbreak, interruption in ecocycles, *etc.* Pesticides adversely affected eco-friendly environmental factors. Among alternative management practice, biological control is found to be safe, economical and permanent. This described as the introduction of natural enemies that cause reduction in insect pest population to keep them below economic threshold level. Effective natural enemies continuously feed on insect pests around the year.

Historically, the term biological control has had various meanings, in the ecological sense; it refers to the 'top-down' action of predators, parasites or pathogens on organisms occupying a lower trophic level, action that maintains their populations at lower level levels than would occur in the absence of these natural enemies, *i.e.* a natural process that does not depend on any human intervention. In the applied sense, it refers to the specific use or manipulation of populations of organisms to achieve reduction or improved regulation of pest populations. In practice, this distinction is not always clear as modern integrated pest management programs usually rest on a foundation of 'background' natural control which occurs without intervention. However, cultural activities may be implemented or modified to facilitate natural biological control, or simply to spare beneficial species from adverse human impacts, including application of pesticides.

The diverse approaches this end is now collectively termed conservation biological control. In contrast, classical biological control involves importing and releasing exotic species with the goal of establishing a self-perpetuating population in a novel geographic region that exerts permanent suppression of a specific pest. Alternatively, augmentation has the goal of immediate, if temporary, reduction of a pest population through occasional or periodic releases of natural enemy. Such releases may be inuntative (large numbers released for immediate effect) or inoculative (fewer individual released with some reproduction expected).

Coccinellidae is a widespread family of small beetles ranging in size from 0.8 to 18 mm (0.03 to 0.71 inches). The family is commonly known as ladybugs in North America, and ladybirds in Britain and other parts of the world. Entomologists widely prefer the names ladybird beetles or lady beetles as these insects are not classified as true bugs.

The majority of coccinellid species are generally considered useful insects, because many species prey on herbivorous homopterans such as aphids or scale insects, which are agricultural pests. Many coccinellids lay their eggs directly in aphid and scale insect colonies in order to ensure their larvae have an immediate food source.

However, some species do have unwelcome effects; among these, the most prominent are of the subfamily Epilachninae (which includes the Mexican bean beetle), which are herbivorous themselves. Usually, epilachnines are only minor agricultural pests, eating the leaves of grain, potatoes, beans, and various other crops, but their numbers can increase explosively in years when their natural enemies, such as parasitoid wasps that attack their eggs, are few. In such situations, they can do major crop damage. They occur in practically all the major crop-producing regions of temperate and tropical countries.

History of Coccinellid Biological Control

Cryptolaemus montrouzieri was the first introduction of predator bio control agent in India in 1898 from Australia to control coffee green scale (*Coccus viridis*) (Puttarudriah *et al.,* 1952; Mayne, 1953). Later, in 1927 ladybird beetles were introduced to control aphids on lentil at Pusa (Dutt, 1927). The pest occupied various parts of country including Mysore, Coorg, Travancore and Bombay. The citrus industry in several states was seriously threatened. A substation for multiplication and liberation of *Rodalia* was set up at Kodaikanal with stocks obtained from Nilgris and New Zealand. Thousands of adults were released for several years in small batches and at intervals, resulting in effective control of the sucking type of pests (Subramanyam, 1955). *C. montotrouzieri* have been reared on the cochineal insect *Dactylopius opuntiae* (Ckll.) in the laboratory and successfully used control *Pulvinaria maxima* (Green) on neem, *Centrococcus insolitus* (Green) on brinjal, *P. psidii* (Mask) and Guava and sapota and *Pseudococcus corymbatus* (Green) on citrus in south India (Rao and David, 1958) and recently against mulberry mealy bug, *Macconellicoccus hirsutus* (Sathe, 1998; Sathe and Bhoje, 2000).

Etimology of Coccinellid

The common name "ladybird" was first used in medieval England, perhaps because these beneficial predators of agricultural pests were believed to be a gift from the Virgin Mary- the "Lady". Other European names have similar associations, such as the German *Marienkafer,* "Marybeetle". This may represent an even older, widespread pre-Christian mythology associated with the beetle.

The scientific name from New Latin *Coccinella,* the diminutive of Latin Coccinus, "scarlet". That from Greek *Kokkinos,* from *kokkos,* "Berry". A related Greek term is *kermes,* an insect that was used to make a scarlet dye. Compare *cochineal,* another red insect and also the bacterium *Coccus,* from the same root.

Role of Coccinellid in Biological Control

Dixon (2000) emphasized three attributes of successful biological control agents: Specificity, Voracity and a high rate of population increase. Whereas these attributes are certainly desirable for rapid impact when the target is increasing in abundance and approaching economic injury levels, many pest populations tend to be transient or seasonally cyclical.

Lady bird beetles, members of coleopteran, well known insects because of their bright colors and beneficial roles. These beetles are also commonly called lady bird beetles, lady bugs and Coccinellids. About 4500 species under 500 genera in 21 tribes are known from the world. In India nearly 350 species under more than 50 genera have been described. Most lady bird species are carnivores. Both adults and grubs are play an important role in checking the population of harmful insects such as aphids, coccids, aleuryds and other soft bodied insects and plant mites. On the other hand another group of coccinellid beetles are phytophagous in habits. There are true vegetarian lady birds that feed on a variety of plants.

They are found in many terrestrial habitats. Presently, there are no endangered ladybird species. Possibly due to their high reproductive capacity and certain protective abilities against of many ladybird species in natural populations are declining because of habitat destruction and use of chemical pesticides.

Human Connections

Lady beetles are a tremendous help to farmers and gardeners, performing natural, nontoxic pest-control. Some of the Asian lady beetle species that were introduced to help control crop pests have become pests themselves, entering houses in foul-smelling masses when the weather turns cold.

Ecosystem Connections

The fertility of aphids and scale insects is staggering, and without legions of tiny predators like lady beetles, lacewings, and others, they have the potential to cause great harm to plants. Unfortunately, indiscriminate forms of pest control also harm insects that are natural exterminators.

Chapter 2

General Morphological Characters of Ladybird Beetle

Ladybird adults are oval, range in length from about 1 mm to over 10 mm depending upon species. Adults of many species have bright red, orange or yellow elytra with distinctive spots, stripes and other markings or patterns. Some species are brightly metallic blue, green or violet.

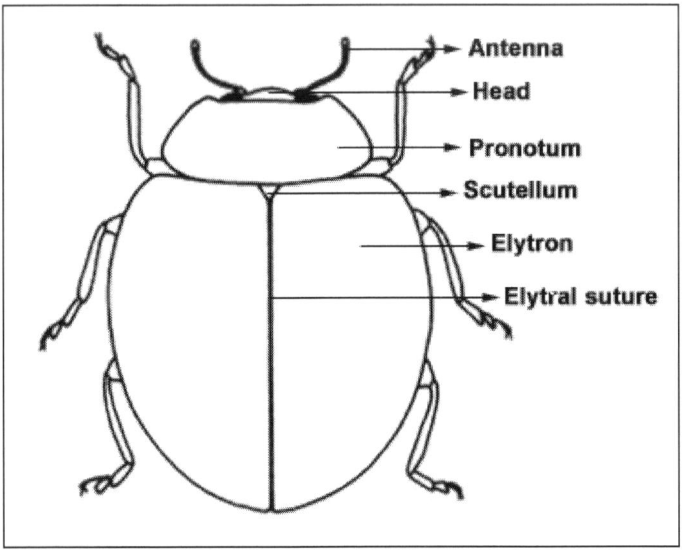

Figure 1: Dorsal View.

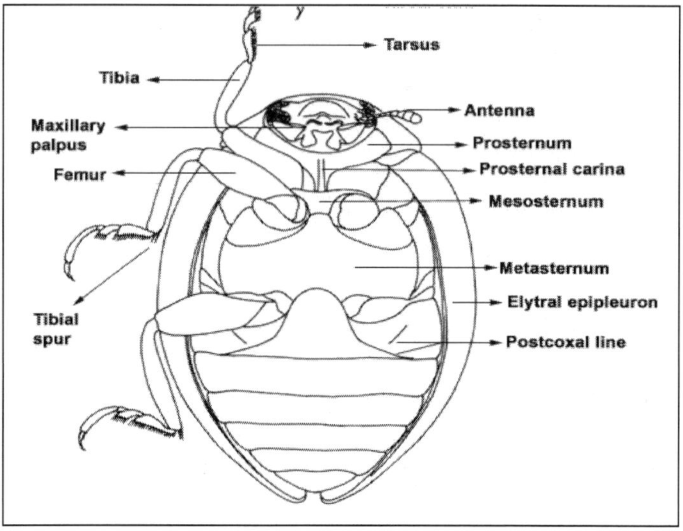

Figure 2: Ventral View.

The body of a Coccinellid is divided into three basic parts: the head, thorax and abdomen. There is triangular scutellum between the pronotum and the elytra. In the dorsal aspect, the pronotum and elytra are broadly and compactly joined. The head is usually retracted into the margin of pronotum at rest. In general, males are smaller than females.

Head

Generally, the head type is prognothus and broader than long and quadrate, with its corners rounded. A feature that is important in the higher clssification of the family is the nature of the clypeus (the portion of the head above the eyes).

The antennae of lady bird beetle are 6-11 segmented. The segments are freely movable. The antennal insertions are usually lateral or dorsolateral, in front of the eyes and broadly separated. The last three or four segments form a loose or compact club in many species.

Compound eyes are well developed and anterolaterally located. They are entire or divided anteriorly by a post-antennal process.

The mouthparts in coccinellids are adapted for bitting and chewing. It consists of labrum, a labium, well developed paired mandibles and maxillae. The mandibles are massive, wide and sickle shaped. The labrum is transverse and quadrate with a truncate base and rounded sides. It is usually narrower than the clyoeus and is visible at the top of the head, just above the anterior apex is simple.

Thorax

The well developed, large dorsal plate of the prothorax, namely called as pronotum. The anterior margin of pronotum in most coccinellids is weakly or strongly and trapezoidally excavated. It is rarely truncate or straight. In some groups, it partially covers the head and is very rarely produced forward to fully cover the head. The scutellum is present between the elytral bases and is usually inversely triangular.

Wings

The hard forewings are called elytra. The elytra has been cover the abdomen and more or less convex dorsally. Its basal margin is nearly straight and the apices are usually rounded and rarely subtruncate. The elytra meet each other at the middle along their entire length forming the sutural line or elytral suture. The hind wings are thin and membranous and form the true functional wings, but are poorly developed or atrophied in some genera (*e.g., Stictobura* Crotch and *Nesolotis* Miyatake) and altogether absent in tribe Lithophilini. Wing polymorphism has been reported in some *Rhizobius* spp. and *Subcoccinella vigintiquatuorpuncata* (L.)

Legs

The legs are well developed and adapted for walking and running. The femora are usually elongate and shallowly grooved on the ventral side to receive the tibiae at rest. The tarsal formula (the no. of tarsal segments in the pro, meso, and meta tarsi) usually expressed as three numbers, (Figure 2b) is of great taxonomic significance and is 3-3-3 or 4-4-4 in coccinellids. The numbers of segments are same for all legs in both male and female. The tarsi are true segments or crypto tetramerous/pseudotrimerous. In the psedudotrimerous or crypto tetramerous condition, (Figure 2a) the tarsi have four segments, but the third is very small, partially hidden by the strongly bilobed second segment. Only tribe Lithophilini (=Tetrabrachini), tarsi are truly tetramerous (Figure 2c).

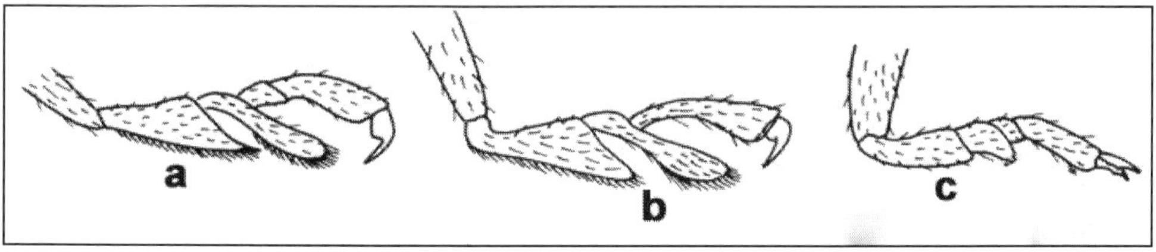

Figure 2: a. Pseudotrimerous/Cryptotetramerous, b. Trimerous, c. Tetramerous.

Abdomen

The abdomen has five or six visible segments on the ventral side. The number of visible segments is usually the same in both sexes and rarely variable. The last visible segments are usually sexually dimorphic. The first abdominal segment usually large in size and has a pair of postcoxal lines, which are characteristic of the family. Their shape, extent and nature of punctuations in the area enclosed by them are important taxonomic characters. The parts of the ninth and tenth abdominal segments are usually defined as the genetalia.

Genetalia

Male Genetalia

The structure of the male genetalia is species specific and is the most important diagnostic character, in the absence of other reliable external diagnostic characters. The male genetalia consist of a tegmen and a siphon. The components of tegmen are a basal piece, a distinct medianprojection or the basal lobe, a pair of lateral arms or parameres and a median basal struct or trabes (Figure 3b).

The tegmen is collectively also referred to as aedeagus. The median lobe is symmetrical or asymmetrical. The parameres are usually setose at their apices and/or outer margins. The sipho is usually elongate, tubular and curved with a basal siphonal capsule and is variously modified at the apex. The siphon is the true penis or the intromittent organ.

Female Genetalia

The female genetalia consist of a pair of genital plates, a spermatheca (receptaculum seminis), a sperm duct and an accesory gland (Figure 3a). A typical spermatheca is differentiated into a nodulus and ramus in the basal part, to which the sperm duct and accessory gland are connected, respectively, and a distal curved arm called cornu. Sometimes, the nodulus and ramus are continuous and not

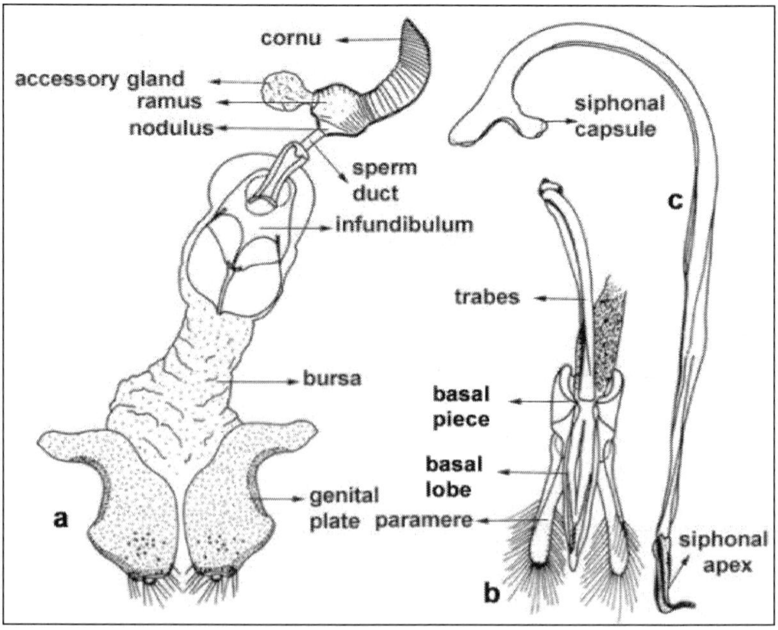

Figure 3: a. Female genetalia; b. Male genetalia.

clearly demarcated. The infundibulum is usually elongate cylindrical; it is highly sclerotized and diagnostically shaped in some groups, but lacking in others. The spermatheca itself is membranous and not visible in some groups (*e.g.*, some genera of Sticholotidini such as *Jauravia, Stichotis* and *Pharoscymnus*).

Chapter 3

Life Cycle of Coccinellids

Lady bird beetles are having complete metamorphosis is called holometabolism and complex metamorphosis, is a term applied to those processes in which the larvae differ markedly from the adults. Insects that undergo holometabolism pass through a larval stage, then enter an inactive stage called pupa, and finally emerge as adults. Coccinellids lay extra infertile eggs with the fertile eggs. These appear to provide backup food source for the larvae when they hatch. The ratio of infertile eggs increases with scarcity of food at the time of egg laying.

Egg

Coccinellidae eggs are 0.2 to 2.0 mm long; white, yellow to red, oval or spindle-shaped and are laid singly or in batches of various sizes, always with long axis perpendicular to the substrate. The chorion is smooth with distinct microsculpture visible only in Epilachninae. When freshly laid, an egg is usually white or creamy-yellow but during embryogenesis it changes colour, becoming darker, often very much darker. This stage lasts 2-18 days. Most Coccinellidae that feed on scale inscets lay their eggs singly or in small groups close to or directly on or beneath the female or immature scale. True ladybird, the Coccinellini are known to form larger batches of up to 100 eggs on leaf surfaces close to the colonies of aphids that are their primary prey. The total number of eggs produced by coccinellids varies considerably between species and is highly dependent on food supply and ambient climatic factors.

Grub

Larvae emerge from eggs either by a rupture at one end of the egg or by a longitudinal split. Shortly before eclosion the chorion becomes thin and

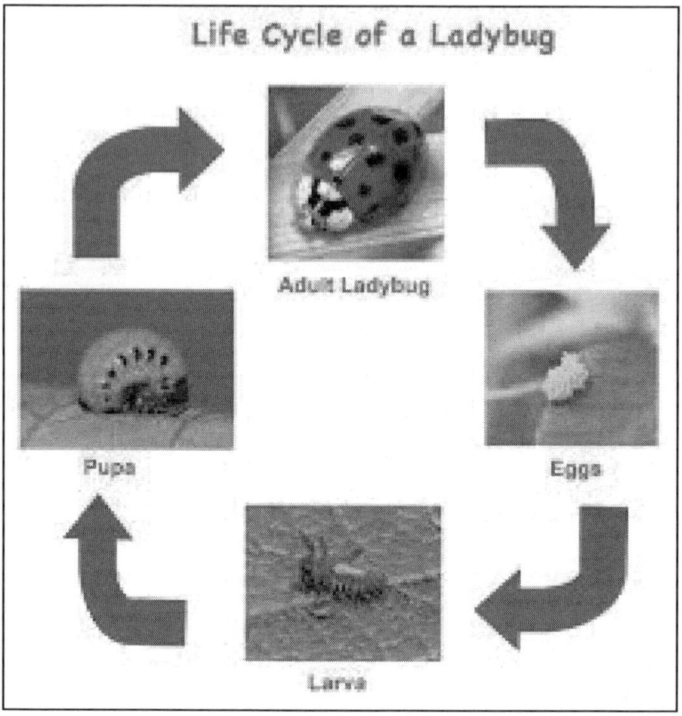

Figure 4: Life Cycle of Coccinellid.

semitransparent, the stemmata, mandibles and hatching spines (egg bursters) on the head of the embryo becoming clearly visible. The freshly hatched larva remains stationary for about an hour for the hardening of its cuticle to take place. Then the larva moves around, often feeding on unhatched eggs (both fertile and unfertile).

There are four stages (instars) in larval life, separated from each other by a moulting period in which the larva casts off its old exoskeleton and builds up a new one, considerably increasing its size. Prior to moulting, the larva stops feeding

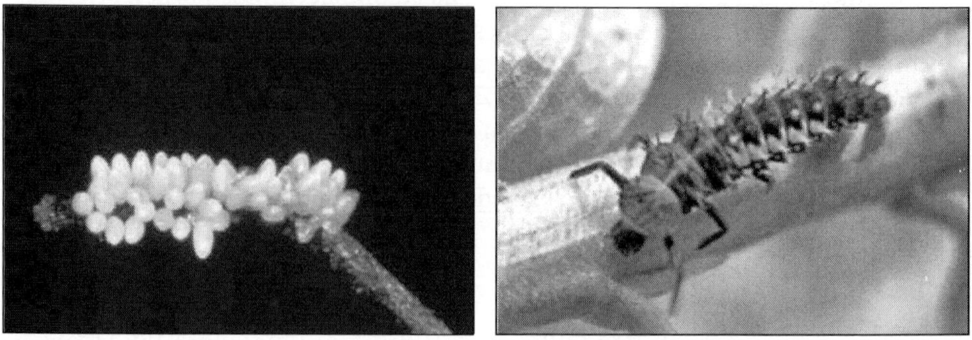

Figure 5: a. Eggs **b. Grub**

and moving and usually attach itself to the substrate using the anal pad (cremaster on the terminal abdominal segment) and remains motionless for several hours. The ecdysis begins at the head and continues along the dorsal part of the thorax and abdomen. The larva liberates itself from the moult and walks away after its cuticle hardens.

The first instar larvae appear to be most restricted in their range of suitable host stages and suffer highest mortality rates. Once their cuticle has hardened the first instar larvae actively seek food. This consists mostly of the eggs of prey in case of coccidophagous species or small aphid nymphs for aphidophaous species. The newly hatched larvae have about one to one and a half days in which to locate their first prey. If unsuccessful in that time, the larva soon dies.

Early authors assumed that all predatory coccinellid larvae use extra oral digestion and that some larvae have perforated or grooved mandibles. In fact coccinellid larval mandibles are not perforated at all. In several species, however, there is a deep groove along the inner surface, used to inject digestive juices into the victim and to suck out juices in cycles until the victim is entirely dry; the prey's empty exoskeleton is then abandoned by the larva.

Coccidophagous species feed on stationary prey that may be ingested as a whole if small or the coccinellid bites the scale covers, sometimes lifting from the substrate first, and then chews the body. Cannibalism occurs frequently among all coccinellids. On emerging, first instar larvae will often feed on unhatched eggs of their own species. Early instar larvae will often feed unhatched eggs of their own species. Early instar larvae and pupae are also preyed upon by later instars and adults of the same species or of other species.

The larvae of coccinellids are of mainly following types:

- ☆ Elongate fusiform: *e.g.*, Coccinellini tribe,
- ☆ Cylindrical oval and more or less convex: *e.g.*, Chilocorini and Epilachnini,
- ☆ Fusiform with wax like exudation: *e.g.*, Scymnini,
- ☆ Elliptical and moderately convex: *e.g.*, Noviini,
- ☆ Short oval and strongly flattened: *e.g.*, Platynaspini and
- ☆ Fusiform and elongate oval: *e.g.*,Sticholotinids

Pupa

Generally, full grown grubs pupate near the point where they lived. However, pupation may also be takes place away from the living place of larvae. Large aggregations of pupae of some species often observed on the host plant of the prey (Figure 6). There are two types of pupae in predaceous coccinellids. In one case, the exuvia is remained at caudal tip and in another case it is nearly intact as noticed in tribe Noviini and Chilocorini.

Figure 6: Pupa of Coccinellid Adult.

The adult beetle emerges from the pupal skin through a slit at the front of the dorsal surface. The body of the tenderal adult is pale and the wings are extended. It takes several hours for the cuticle to harden and to develop the characteristic colour pattern and freshly emerged specimens remain distinctly lighter than specimens from the previous generation for quite some time. Adults mate within a few days after emergence, and oviposition follows from 5 to 25 days after eclosion. Copulation typically lasts for 15-60 minutes but can extend for few hours or even days. Longevity in Coccinellidae is related to prey synchrony and dormancy mechanisms. Species in which the adults undergo long periods of aestivation may live up for two years. However, multivoltine species in warmer climates usually only live up to two months.

Figure 7: Adult Ladybird Beetle.

Chapter 4

Mating and Reproduction

Morphologically, ladybirds could probably be sexed owing to subtle patterns on the body (Omkar and Pervez, 2000; Omkar and Pathak, 2007). However, this dimorphism is not common in ladybirds. Nevertheless, they have a size-dependent sexual dimorphism with females being larger than males. The males are consistently smaller in body size than females despite of the similar developmental period (Dixon, 2000). Male body size is a trade-off to its gonadal requirements needed for the sperm production and sexual activity (Yasuda and Dixon, 2002). Females have a larger body size for a quantitative egg production and accommodation, as larger females have more ovarioles in their gonads than smaller females (Dixon and Guo, 1993). Smaller males could be better mates in the limited food scenario and will spend less time eating and more time mating.

Mating

Mostly, ladybirds are protandrous, *i.e.*, males mature earlier than females and males exhibit subtle courtship (approach, watch, examine, embrace, and mount) before entering into mating (Omkar and Pervez, 2005). Certain coccidophagous female ladybirds are choosy and prefer previously-mated males than virgins (Jiaqin *et al.*, 2014). Choosy females possibly identify previously-mated males by marking them with their cuticular hydrocarbons during copulations. Many polymorphic aphidophagous female ladybirds preferentially select melanic and darker morphs as potential mates, especially during winter (Srivastava and Omkar, 2005; Mishra and Omkar, 2014; the choice for a particular morph in certain ladybirds is also dependent on the seasonal variations (Wang *et al.*, 2009). This choice has a direct significant effect on their reproductive performance. Many ladybird species are promiscuous and females mate multiple times (Srivastava and Omkar, 2005). This occurs even when sperm is not limited. Females benefit from sperm provided by multiple mates.

In laboratory conditions, polyandry benefits females for both quantitative progeny and paternity (Haddrill *et al.*, 2007). The apparent reproductive benefits of polyandry with mate choice in ladybirds are more likely a result of additive fecundity stimulation as a function of multiple matings, rather than a result of females choosing better males (Pervez and Maurice, 2011; Bayoumy and Michaud, 2014). Interspecific matings are also conspicuous in ladybirds (Noriyuki *et al.*, 2012).

Copulation is prolonged if heavier adults mate compared with lighter adults with increased paternity (Pervez and Singh, 2013). This duration-dependent higher paternity for males can possibly be ascribed to the transfer of larger numbers of sperm in multiple spermatophores (Haddrill *et al.*, 2008). Multiple matings benefit ladybirds (Omkar and Srivastava, 2002; Omkar and James, 2005; Omkar and Pervez, 2005; Omkar *et al.*, 2010; Bista and Omkar, 2012) especially polyandry and promiscuity (Omkar and Mishra, 2005).

Figure 8: Mating of Ladybird.

The sperm transfer mechanism is both direct (Omkar *et al.*, 2006) and indirect via a spermatophore, which could also be proposed as a "nuptial gift" as most females devour it after copulation (Perry and Rowe, 2010). However, mating and spermatophore transfer is costly in some ladybirds as there is a subtle trade-off between mating and the longevity (Mishra and Omkar, 2006). The spermatophore costs involve a trade-off between current and future reproductive investment, as the efficiency of sperm transfer decreases or halted in successive matings (Perry and Tse, 2013). Males ingest more food following spermatophore transfer indicating that spermatophore transfer depletes male nutritional reserves and has postmating implications. Nevertheless, in directly inseminating ladybirds without a spermatophore, food depletion is compensated by producing higher relative sperm content in the ejaculate (Perry and Rowe, 2010).

Reproduction

Adult body size has a prominent effect on reproduction in ladybirds. As mentioned above, heavier males mate for a longer duration, which is positively correlated with oviposition period, fecundity, and fertility in predaceous ladybirds (Pervez and Singh, 2013). As ladybirds have a direct sperm transfer mechanism, they can produce fertile eggs even after 1min of copulation, indicating that the sperm transfer mechanism is initiated very early in the copulation act (Omkar *et al.*, 2006). However, these ladybirds mate for a few hours on a single mating event, which suggests that sperm transfer mechanism is early and continuous. Ladybirds have multiple ejaculations during a single complete mating which accounts for their increased fecundity and fertility. While an increase in the number of matings leads to increased egg production, it has a reproductive cost in terms of longevity (Omkar and Mishra, 2005).

Age, Aging, and Rhythmicity

Both paternal and maternal age influence mating performance and reproductive outcome. It is likely that maternal age affects fecundity, oviposition rate, and the reproductive and nonreproductive periods and paternal age affects percent egg viability. Nonetheless, parental age per se has an indirect implication on these aspects probably due to age-influenced variation in seminal proteins that affects both oviposition and fecundity (Omkar *et al.*, 2010). However, paternal effects diminish as a function of mating history, suggesting that transgenerational signals of male origin are also subject to depletion (Michaud *et al.*, 2013). Ladybirds have sexdependent aging trajectories with reproductive performance declining later in females than in males (Mishra and Omkar, 2006). These trajectories are unsynchronized with sex; female virility diminishes later than that of males. Optimization of age difference between mates maximizes the reproductive output; females that are five days older than the males during early life produce better progeny in quantitative terms (Omkar and Mishra, 2009). This was the first study on insects.

Chapter 5

Classification and Ladybird Species Distribution in South and East Asia

Lady bugs, ladybirds, ladybird beetles, lady beetles which are just common terms for the beloved, familiar coccinellidae family of beetles. Ladybird beetles have quite an impressive taxonomic classification. They belong to the largest phylum, the largest class and the largest order. The coccinellidae are markably sizable family as well.

History of Coccinellid Classification

No doubt because of their common occurrence in all over the world and distinctiveness, Linnaeus in 1758 established the genus Coccinella and described 36 species that currently are assigned to several genera and some to different beetle families. However, it was about 100 years before two major works by the French entomologist, Etienne Mulsant (1848 and 1850), set the foundations for the modern classification of Coccinellidae. Mulsant's publications included first ever supra-generic categories (Tribes) in Coccinellidae, keys to genera and descriptions of all known world species.

Crotch (1874) provided a critical catalogue of the world taxa and proposed a slightly altered classification that departed from the division of Coccinellidae into the hairy 'Trichoisomides' and glabrous 'Gymnosomides' of Mulsant and classified setose Ephilachnides as a subgroup of the subfamily and synonymised many of the Mulsant taxa.

According to Gordon (1985), Weise was apparently the first coccinellid taxonomist to realize that male genetalia could be useful to distinguish species.

Certainly the publication of 'Phylogeny of the family Coccinellidae (Coleoptera)' by H. Sasaji in 1968 was a landmark in coccinellid research. He studied and analyzed critically both adult and larval characters of Coccinellidae.

Coccinellidae are a family of beetles belonging to the super family Cucujoidea, which in turn belongs to the series Cucujiformia within the suborder Polyphaga of the beetles (Coleoptera). It is divided into six sub families: Coccidulinae, Coccinellinae, Scymninae, Chilocorinae, Sticholotidinae and Epilachninae (Vandenberg 2002). Except for the mychophagousCoccinellinae (*Halyziini and Tythaspis*) and the phytophagousEpilachninae, all remaining coccinellids are predators of hemipteran insects.

Sticholotidinae

Terminal segment of maxillary palpi elongate oval or elongate; mentum very narrowly articulated with submentum; the 9thsternite of male flat, triangular; head capsules sometimes projecting anteriorly. Antennae inserted usually more or less dorsally and rarely laterally; mandibles simple or bifid at apex; inner process of hypomera of prothorax relatively short (anterior coxal cavities broadly open behind); middle coxal cavities broadly separated by compact and broad articulation of meso and metasterna; tarsi usually cryptotetramerous and rarely trimerous; coxite of female very elongate; tegmen of male genetalia with elongate process at its basal base, siphon relatively weakly curved and without distinct capsule. Body minute or small in size.

Chilocorinae

Clypeus strongly expanded laterally; anterior margin of pronotum deeply and trapezoidally concave and lateral portions of pronotum very strongly descending below; elytral base distinctly broader than pronotal base; metasternum distinctly impressed for the middle femora; elytral epipleura relatively broad and or its inner carina reaching elytral apex; tibiae often angulated externally.

Scymninae

Clypeus not strongly expanded laterally; anterior margin of pronotum not deeply or not trapezoidally concave, and lateral portions descending below but not very strongly; elytral base slightly broader than pronotal base; elytral epipleura very narrow and its inner carina not reaching the elytral apex but ending near two thirds of elytral length from base.

Coccinellinae

Dorsum glabrous; mandible with a bifid or multidenticulate tip and also with a basal tooth; mesoepimeron nearly triangular, its posterior margin almost straight or slightly angulate, mentum relatively narrowly articulated with submentum and distinctly divergent apically; inner margin of coxite of female genitalia without a small emargination and with a stylus being near the inner corner of coxites; siphon of male genitalia rather long and usually strongly curved with a well-developed capsule, median piece of tegmen usually not very slender.

Epilachninae

Dorsum pubescent; mandible with a multidendiculate tip and without a basal tooth; mesoepimeron quadrate, its posterior margin distinctly angulate; mentum very broadly articulated with submentum and convergent apically; maxillary galea broad and quadrate in shape and bearing numerous thin hairs; antennal insertions more between than before the eyes; anterior margin of clypeus truncate; inner margin of coxite of female genitalia usual with a small emargination and a stylus being at far outward from the inner corner of coxite; siphon of male genetalia rather than short and weakly curved without a well-developed capsule, median piece of tegmen tubular and slender.

Ladybird Species (Only Predators) Distribution Around the World

Subfamily: Chilocorinae

Tribe: Chilocorini

1. *Brumoides lineatus* (Weise)

   Distribution: India (AN), Sri Lanka, Bangladesh, Nepal, Thailand and Myanmar.

2. *B. suturalis* (Fabricus)

   Distribution: India (GA, KA, TN, PB, MH, UP, MA), Pakistan, Bangladesh, Sri Lanka, Bhutan, Nepal and Sri Lanka.

3. *Brumus ceylonicus* (Weise)

   Distribution: Sri Lanka.

4. *Chilocorus bipustulatus* (Linnaeus)

   Distribution: India (J&K), Pakistan, China, Mongolia, Western Europe and Russia.

5. *C. braeti* (Weise)

   Distribution: India (WB, SK).

6. *C. circumdatus* (Gyllenhall)

   Distribution: India (AS, KA, TN), Sri Lanka, Nepal, Indonesia, China, Hong Kong, Introduced into Australia.

7. *C. coelosimilis* (Kapur)

   Distribution: India (AN, UP, GA).

8. *C. hauseri* (Weise)

   Distribution: India (WB, SK), Myanmar and China.

9. *C. infernalis* Mulsant)

   Distribution: India (ME, HP, J&K, UP), Nepal, Pakistan and China.

10. *C. matsumurai* (Miyatake)

    Distribution: Nepal.

11. *C. melanophthalmus* (Mulsant)

    Distribution: Indonesia (Java).

12. *C. melas* (Weise)

    Distribution: India (SK), Bhutan, Myanmar, China, Hong Kong and Southeast Asia.

13. *C. nigrita* (Fabricus)

    Distribution: India (KA, KL, AP, TN, MH), Pakistan, Sri Lanka, Bangladesh, Myanmar, China, Indonesia, Thailand, South Africa, Seychelles, Pacific and Brazil.

14. *C. politus* (Mulsant)

    Distribution: India, Bhutan, Nepal, Thailand and Java (Indonesia).

15. *C. rubidus* (Hope)

    Distribution: India (GA, J&K, UP), Pakistan, Nepal, China, Malaysia, Japan, Celebes Australia, Korea, Siberia, Manchuria and Mongolia.

16. *C. subindicus* (Booth)

    Distribution: India (LDI, MH, TN, KL) Maldives and Sri Lanka.

17. *Curinus coerueus* (Mulsant)

    Distribution: India (KA, TN, AP), Nepal, Thailand, Philippines, Central and South America.

18. *Exochomus nigripennis* (Erichson)

    Distribution: North western India, Pakistan, Palaearctic and Africa,

19. *E. pubescens* (Kuster)

    Distribution: India (Delhi (Kapur, 1951a), Pakistan, Spain, North Africa, Greece, Egypt, Syria and Palestine.

20. *E. quadripustulatus* (Liinaeus)

    Distribution: India (J&K, Middle East), Russian Far East and Western Europe.

21. *Orcus bipunctatus* (Gorham)

    Distribution: Myanmar.

22. *O. carinicus* (Gorham)

    Distribution: Myanmar.

23. *O. quadriguttatus* (Gorham)

    Distribution: Myanmar.

24. *Phaenochilus flaviceps* (Miyatake)
    Distribution: India (TN).
25. *P. indicus* (Miyatake)
    Distribution: India (TN).
26. *Priscibrumus disjunctus* (Canepari)
    Distribution: Nepal.
27. *P. himalayensis* (Kapur)
    Distribution: Nepal.
28. *P. lituratus* (Gorham)
    Distribution: India (HP, J&K) and Pakistan.
29. *P. trijunctus* (Kapur)
    Distribution: India (J&K, Karakoram Range).
30. *P. trubetzkoi* (Barovsky)
    Distribution: India (J&K).
31. *P. uropygialis* (Mulsant)
    Distribution: India (J&K, HP), Bhutan, Pakistan and Nepal.
32. *Simmondsius pakistanensis* (Ahmad and Ghani)
    Distribution: India, Pakistan.

Tribe Platynaspidini

33. *Phymatosternus khalaus* (Bielawski)
    Distribution: Bhutan.
34. *P. lewisi* (Crotch)
    Distribution: India, Myanmar, China, Japan and Taiwan.
35. *P. octoguttatus* (Miyatake)
    Distribution: Bhutan and China.
36. *P. samchus* (Bielawski)
    Distribution: Bhutan.
37. *P. wittmeri* (Bielawski)
    Distribution: Bhutan.
38. *Platynaspidius angulimaculatus* (Mader)
    Distribution: Nepal, Thailand and China.
39. *P. bimaculata* (Pang and Mao)
    Distribution: India (North Eastern States), China.
40. *P. saundersi* (Crotch)
    Distribution: India (HP, J&K, WB), Nepal and Afghanistan.

41. *Platynapsis flavoguttata* (Gorham)
    Distribution: India (KA (Belgaum) and Myanmar.
42. *P. nepalensis* (Canepari)
    Distribution: Nepal.
43. *P. ocellimaculata* (Pang and Mao)
    Distribution: Nepal and China.
44. *P. trimaculata* (Weise)
    Distribution: India (SK).

Tribe: Telsimiini

45. *Telsimia bangalorensis* (Kapur)
    Distribution: India (KA).
46. *T. bicolor* (Kapur)
    Distribution: India (AN, Great Nicobar).
47. *T. ceylonica* (Weise)
    Distribution: India (KA) and Sri Lanka.
48. *T. darjeelingensis* (Kapur)
    Distribution: India (WB).
49. *T. flavomaculata* (Poorani)
    Distribution: India (KA).
50. *T. humidiphila* (Kapur)
    Distribution: India (ME) (Cherrapunji).
51. *T. martis* (Mulsant)
    Distribution: "India Orientali".
52. *T. postocula* (Kapur)
    Distribution: India (AN).
53. *T. rotundata* (Motschulsky)
    Distribution: Sri Lanka, Philippines and Taiwan.

Subfamily: Coccidulinae

Tribe: Coccidulini

54. *Empia vittata* (Weise)
    Distribution: Srilanka.
55. *Oridia pubescens* (Gorham)
    Distribution: India (SK, WB) and Myanmar.

Tribe: Noviini

56. *Rodalia amabilis* (Kapur)
    Distribution: India (KA, KL, TN, WB, OR) and Sri Lanka,
57. *R. andamanica* (Weise)
    Distribution: India (AN).
58. *R. breviuscula* (Weise)
    Distribution: India (TN, KA, KL, BR, WB), Sri Lanka, Myanmar, Thailand and China.
59. *R. cardinalis* (Mulsant)
    Distribution: India (KA, TN, MH), Sri Lanka, Australia and Nearctic.
60. *R. fulvescens* (Hoang)
    Distribution: India (AN), Bangladesh and Vietnam.
61. *R. fumida* (Mulsant)
    Distribution: India (AS, MA, WB, PB, HP, ME, BR, MP, DL, KA, UP, MH, TN) and Pakistan.
62. *R. marginata* (Bielawski)
    Distribution: Java and Nepal.
63. *R. minima* (Kapur)
    Distribution: India (TN, KA).
64. *R. netara* (Kapur)
    Distribution: India (TN).
65. *R. nigrofrontalis* (Kapur)
    Distribution: India (KL).
66. *R. octoguttat* (Weise)
    Distribution: India (UP, ME), Nepal, Sri Lanka and Myanmar.
67. *R. ruficollis* (Mulsant)
    Distribution: India (AS, TN, WB), Pakistan, Thailand.
68. *R. sexotata* (Mulsant)
    Distribution: India (PY, SK, TN, WB, UP), Nepal, Hong Kong and Bhutan.

Tribe: Sumniini

69. *Sumnius cardoni* (Weise)
    Distribution: India (BR, UP, OR, SK, AS), Pakistan, China and Thailand.
70. *S. haematica* (Gorham)
    Distribution: India (AS, Himalayas) and Bhutan.
71. *S. hoangi* (Canepari)
    Distribution: Nepal.

72. *S. notivestis* (Mulsant)

    Distribution India (BR), Pakistan.

73. *S. thanhhoaensis* (Hoang)

    Distribution: Nepal. Vietnam.

74. *S. vestita* (Mulsant)

    Distribution: India (UP, BR, WB), Pakistan, Bhutan, Nepal and Sri Lanka.

Tribe: Tetrabrachini

75. *Tetrabrachys gandhara* (Kapur)

    Distribution: Pakistan.

76. *T. lindemannae* (Fursch)

    Distribution: Pakistan and Nepal.

77. *T. robusta* (Kapur)

    Distribution: India (UP) and Pakistan.

Subfamily: Coccinellinae

Tribe: Bulaeini

78. *Bulaea lichatschovii* (Hummel)

    Distribution: India (RJ, UP), Central and West Asia, Afghanistan, Mediterranean region, North and Central Africa.

79. *B. nevilli* (Dohrn)

    Distribution: Nepal.

80. *Adalia bipunctata* (Linnaeus)

    Distribution: Pakistan, Middle East, China, Africa, Europe, North America, South America and Australia.

81. *A. puetzi* (Iablokoff-Khnzorian)

    Distribution: India (UP) and Pakistan.

82. *A. simmondsi* (Kapur and Sudha Rao)

    Distribution: India (HP).

83. *A. tetraspilota* (Hope)

    Distribution: India (J&K, WB, HP, UP), Nepal, Iran, Afghanistan, Central Asia and China.

84. *Aiolocaria hexaspilota* (Hope)

    Distribution: India (J&K, SK, WB), Pakistan, Nepal and Myanmar.

85. *Alloneda dodecaspilota* (Hope)

    Distribution: India (SK, WB), Nepal, Myanmar, Vietnam, Thailand and China.

86. *Anatis ocellata* (Linnaeus)
    Distribution: Pakistan, China, Korean Peninsula and Europe.
87. *A. quindecimpunctata* (De geer)
    Distribution: Pakistan.
88. *Anegleis cardoni* (Weise)
    Distribution: India (KL, TN, MP, UP, KA, BR, GA), Pakistan and Sri Lanka.
89. *Bothrocalvia albolineata* (Gyllenhall)
    Distribution: India (ME, SK, WB), China, Hong Kong and Taiwan.
90. *B. lewisii* (Crotch)
    Distribution: India, China and Myanmar.
91. *B. pupillata* (Swartz)
    Distribution: India (WB), Nepal and Java.
92. *Callicaria superba* (Mulsant)
    Distribution: India (HP, SK, WB, UP), Nepal, Bhutan, Tibet, Japan and Taiwan.
93. *Calvia albida* (Bielawski)
    Distribution: India (SK, WB, UP) and Nepal.
94. *C. andrewesi* (Weise)
    Distribution: India (TN, KL).
95. *C. breiti* (Mader)
    Distribution: India (HP, J&K, SK, UP), Nepal, Pakistan and Bhutan.
96. *C. championorum* (Booth)
    Distribution: India (UP), China and Taiwan.
97. *C. flaveola* (Booth)
    Distribution: India (UP, J&K), Pakistan and Myanmar.
98. *C. monosha* (Bielawski)
    Distribution: Bhutan.
99. *C. punctata* (Mulsant)
    Distribution: India (UP, J&K), Nepal, Pakistan and Afghanistan.
100. *C. quatuordecimguttata* (Linnaeus)
    Distribution: India (WB), Bhutan and Nepal.
101. *C. quidecimguttata* (Fabricus)
    Distribution: India (Kuznetsov, 1997), Japan, China, Mongolia and Europe.
102. *C. shiva* (Kapur)
    Distribution: India (SK, MA, UP) and Nepal

103. *C. sykesii* (Crotch)

Distribution: India (AS, WB, ME, TN, Deccan) and Nepal.

104. *C. tricolor* (Korschefsky)

Distribution: India (WB, ME).

105. *C. vulnerata* (Hope)

Distribution: India (SK, WB, UP, DL) and Nepal.

106. *Cheilomenes sexmaculata* (Fabricus)

Distribution: Throughout India, Bangladesh, Pakistan, Sri Lanka, Bhutan, Myanmar, Malaysia, Indonesia, Philippines, Vietnam, China, Japan and Australia.

107. *Coccinella lama* (Kapur)

Distribution: India. Tibet.

108. *C. luteopicta* (Mulsant)

Distribution: India (SK, WB, UP), Nepal, Tibet, China.

109. *C. magnopunctata* (Rybakow)

Distribution: India (J&K), Mongolia, Nepal, China and Tibet.

110. *C. marussii* (Kapur)

Distribution: India (J&K, Karakoram Range).

111. *C. nepalensis* (Iablokoff-Khnzorian)

Distribution: Nepal.

112. *C. nigrovittata* (Kapur)

Distribution: India, Tibet, China, Russia and Mongolia.

113. *C. saucerottei* (Mulsant)

Distribution: India (J&K), Tibet and Palaearctic.

114. *C. septumpunctata* (Linnaeus)

Distribution: All over India, Pakistan, Sri Lanka, Palaearctic and North America.

115. *C. transversalis* (Fabricus)

Distribution: New Zealand.

116. *C. transversoguttata* (Faldermann)

Distribution: India (ME), China, Japan, Siberia and North America.

117. *C. undecimpunctata* (Linnaeus)

Distribution: India (J&K, UP), Nepal, Pakistan, China, Mongolia, North Africa, North America, Europe and Australia.

118. *C. undecimpunctata aegyptiaca* (Reiche)

Distribution: India

119. *Coelophora biplagiata* (Swartz)

Distribution: India, Nepal, Myanmar, Indochina, China, Korea, Japan, Taiwan and Indonesia.

120. *C. bissellata* (Mulsant)

Distribution: India (KA, TN, AS, MA, SK, UP, ME, HP, KL, WB), Nepal, Bangladesh, China, Thailand, Vietnam, Sumatra, Philippines and New Guinea.

121. *C. bowringii* (Crotch)

Distribution: India (AS) and Myanmar.

122. *C. circumusta* (Mulsant)

Distribution: India (TN, J&K), Nepal, Thailand, South China and Taiwan.

123. *C. circumvelata* (Mulsant)

Distribution: Nepal, The Philippines.

124. *C. decemguttata* (Weise)

Distribution: India (SK).

125. *C. duvaucelii* (Mulsant)

Distribution: India, Myanmar, China and Indonesia.

126. *C. inaequalis* (Fabricus)

Distribution: India, Sri Lanka, Southeast Asia and Australia.

127. *C. minki* (Mulsant)

Distribution: Sri Lanka and Java.

128. *C. newporti* (Mulsant)

Distribution: Myanmar and Philippines.

129. *C. nitidicollis* (Kapur)

Distribution: India (SK).

130. *C. octosignata* (Mulsant)

Distribution: India, Indonesia.

131. *C. saucia* (Mulsant)

Distribution: India (WB), Nepal, Thailand, China, Taiwan and Philippines.

132. *Eoadalia juliae* (Mulsant)

Distribution: India.

133. *Harmonia axyridis* (Pallas)

Distribution: Nepal, China, Taiwan, Japan and Palaearctic.

134. *H. dimidiata* (Fabricus)

Distribution: India (AS, HP, RJ, PB, SK, J&K, WB), Pakistan, Nepal, Bhutan, China, Japan, Taiwan and Introduced into North America.

135. *H. dunlopi* (Crotch)

  Distribution: India. Malaysia, Philippines.

136. *H. eucharis* (Mulsant)

  Distribution: India (HP, SK, UP), Pakistan, Myanmar, Nepal, Himalayas and Southern China.

137. *H. expallida* (Weise)

  Distribution: India (J&K,HP).

138. *H. octomaculata* (Fabricus)

  Distribution: Almost throughout India, Pakistan, Nepal, Bangladesh, Sri Lanka, Widespread in the Orient, Micronesia and Australia.

139. *H. sedecimnotata* (Fabricus)

  Distribution: India (WB, SK) Nepal and Southeast Asia.

140. *Heteroneda billardieri* (Crotch)

  Distribution: Myanmar, Indonesia and China.

141. *Hippodamia arctica* (Schneider)

  Distribution: Nepal, China and Palaearctic.

142. *H. heydeni* (Weise)

  Distribution: India (J&K) and Tibet.

143. *H. tredecimpunctata* (Linnaeus)

  Distribution: India (HP, Northwestern region), Nepal, Tibet and Palaearctic.

144. *H. variegate* (Goeze)

  Distribution: India (J&K, UP, HP, MH), Nepal, Pakistan, Afghanistan, Tibet, Mongolia, China, Northern and eastern Africa and Palaearctic.

145. *H. variegate doubledayi* (Mulsant)

  Distribution: North India.

146. *Illeis bistigmosa* (Mulsant)

  Distribution: India (KA), Sri Lanka, Nepal, Thailand, Indonesia, Malaysia, Philippines and Celebes.

147. *I. cincta* (Fabricus)

  Distribution: India, Sri Lanka and Indonesia.

148. *I. confuse* (Timberlake)

  Distribution: India (AS), Nepal, Thailand, China and Hong Kong.

149. *I. indica* (Timberlake)

  Distribution: India (UP, J&K), Pakistan and Thailand.

150. *I. kapuri* (Anand *et al.*)

  Distribution: North India.

151. *I. koebelei* (Timberlake)

 Distribution: India (AS), Japan, Taiwan and China.

152. *Megalocaria dilatata* (Fabricus)

 Distribution: India (HP, WB, AS, KA, TN), Sri Lanka, Nepal, Malaysia and Guam.

153. *M. pearsoni* (Crotch)

 Distribution: India.

154. *Micraspis allardi* (Mulsant)

 Distribution: India (BR, UP, PB), Nepal, Pakistan, Myanmar and Indonesia.

155. *M. discolor* (Fabricus)

 Distribution: India, Bangladesh, Sri Lanka, Pakistan, Almost throughout the Orient and Palaearctic.

156. *M. guerini* (Mulsant)

 Distribution: India (TN).

157. *M. univittata* (Hope)

 Distribution: India (TN, KA, AP) and Nepal.

158. *M. vincta* (Gorham)

 Distribution: India (AS, BR, TN, AP, KA), Nepal, Bangladesh, Myanmar and Thailand.

159. *M. yasumatsui* (Sasaji)

 Distribution: India (WB, OR) and Bangladesh.

160. *Oenopia adelgivora* (Poorani)

 Distribution: India: SK.

161. *O. billieti* (Mulsant)

 Distribution: India (AS, J&K, PB, UP, KA, SK, ME, HP), Nepal and Tibet.

162. *O. conglobata* (Linnaeus)

 Distribution: India (J&K, HP), Pakistan, China, Mongolia, Europe, North Africa and North America.

163. *O. diabolica* (Canepari)

 Distribution: Nepal.

164. *O. kirbyi* (Mulsant)

 Distribution: India (ME, WB, MA, J&K, UP, SK), China, Thailand and Myanmar.

165. *O. mimica* (Weise)

 Distribution: India (SK, HP, UP) and Nepal.

166. *O. quadripunctata* (Kapur)

 Distribution: India (ME, WB, SK, AS, MH), Bhutan and Myanmar.

167. *O. sauzeti* (Mulsant)

Distribution: India (MA, SK, UP, WB, HP, AS, J&K) Bhutan, Pakistan, Nepal, Myanmar, Thailand and China.

168. *O. sexareata* (Mulsant)

Distribution: India (MH, WB, TN, BR, AS, UP, SK), Bhutan, Nepal, Myanmar, China, Vietnam and Indonesia (Sumatra).

169. *O. signatella* (Mulsant)

Distribution: India (WB, SK, ME, MH), Bhutan, Myanmar and Nepal.

170. *O. smetanai* (Canepari)

Distribution: India (UP) and Nepal.

171. *O. walteri* (Sicard)

Distribution: India (TN).

172. *Palaeoneda auriculata* (Mulsant)

Distribution: India (HP) and Nepal.

173. *Phrynocaria congener* (Billberg)

Distribution: India, China and Japan.

174. *P. eberti* (Bielawski)

Distribution: Nepal.

175. *P. funebris* (Crotch)

Distribution: India (KA).

176. *P. perrotteti* (Mulsant)

Distribution: India (PY, TN, KA, MP, OR, PB).

177. *P. unicolor* (Fabricus)

Distribution: India (AN, OR, WB, MP).

178. *P. vidua* (Mulsant)

Distribution: India, Indonesia, Philippines, New Guinea and Taiwan.

179. *Propylea dissecta* (Mulsant)

Distribution: India (AS, TN, KA, UP) Bangladesh and Nepal.

180. *P. japonica* (Thunberg)

Distribution: India, Japan, Korea, China and Taiwan.

181. *P. luteopustulata* (Mulsant)

Distribution: India (AS, WB, SK, ME, UP, HP, AN), Nepal, Bhutan, Himalayas, Sri Lanka, Myanmar, Vietnam, Thailand, Tibet and China.

182. *P. quatuordecimpunctata* (Linnaeus)

Distribution: India, Pakistan, Bangladesh, Japan, China, Europe and North America.

183. *Synona martini* (Lablokoff-khnzorian)
Distribution: India and TN.
184. *S. melanaria* (Mulsant)
Distribution: India.
185. *S. melanopepla* (Mulsant)
Distribution: India.
186. *S. rougeti* (Mulsant)
Distribution: India (KA, KL, TN, WB, UP), Sri Lanka, Bhutan, China, Taiwan, Vietnam and Philippines.
187. *Synonycha grandis* (Thunberg)
Distribution: India (AN, KA, TN, WB, SK, MH), Sri Lanka, Nepal, Japan, Indonesia and China

Tribe: Psylloborini

188. *Halyzia feae* (Gorham)
Distribution: Myanmar and Bengal.
189. *H. nepalensis* (Canepari)
Distribution: Nepal and Himalayas.
190. *H. sanscrita* (Mulsant)
Distribution: India (SK, UP), Nepal, China and Tibet.
191. *H. straminea* (Hope)
Distribution: India (SK, WB, ME, HP, AN), Nepal, Bhutan and China.
192. *H. tschitscherini* (Semenow)
Distribution: India (HP), Afghanistan, Tibet (Pamir) and Turkestan.
193. *Macroilleis hauseri* (Mader)
Distribution: India (WB), Bhutan, China and Taiwan.
194. *Protothea quadripunctata* (Mulsant)
Distribution: India (SK).
195. *Psyllobora bisoctonotata* (Mulsant)
Distribution: India (PB, UP, KA) and Pakistan.
196. *Vibidia korschefskii* (Mader)
Distribution: India and China.

Tribe: Singhikaliini

197. *Singhikalia ornate* (Kapur)
Distribution: India (SK), Vietnam and Taiwan.

Subfamily: Epilachninae

Tribe: Epilachnini

198. *Afidenta misera* (Weise)

 Distribution: India, Sri Lanka, Nepal, China, Taiwan, Vietnam, Philippines and Indonesia.

199. *Afidentula bisquadripunctata* (Gyllenhal)

 Distribution: India (TN), Vietnam, China, Nepal and Sri Lanka.

200. *A. erberi* (Fursch)

 Distribution: India.

201. *A. herbigrada* (Mulsant)

 Distribution: India and Sri Lanka.

202. *A. himalayana* (Kapur)

 Distribution: India (SK, UP, WB).

203. *A. manderstjernae* (Mulsant)

 Distribution: India, Nepal, Vietnam and China.

204. *A. minima* (Gorham)

 Distribution: India (KA).

205. *A. stephensi* (Mulsant)

 Distribution: India and Nepal.

206. *Afissula antennata* (Bielawski)

 Distribution: Burma.

207. *A. burmana* (Bielawski)

 Distribution: Burma.

208. *A. hyla* (Bielawski)

 Distribution: Burma.

209. *A. kambaitana* (Bielawski)

 Distribution: Nepal. China. Myanmar.

210. *A. merkli* (Jadwiszczak)

 Distribution: India: WB.

211. *A. mysticoides* (Sicard)

 Distribution: India. Nepal. China.

212. *A. parvula* (Crotch)

 Distribution: India and Nepal.

213. *A. puncta* (Bielawski)

 Distribution: Burma.

214. *A. rana* (Kapur)
Distribution: Nepal and China.
215. *A. sadonensis* (Bielawski)
Distribution: Myanmar.
216. *A. sanscrita* (Crotch)
Distribution: India (SK) and China.
217. *Epilachna andrewesi* (Gorham)
Distribution: India (TN: Nilgiri Hills).
218. *E. atypical* (Dieke)
Distribution: India (WB) and Indonesia.
219. *E. bengalica* (Dieke)
Distribution: India (SK, WB), Nepal and Bhutan.
220. *E. besuheti* (Canepari)
Distribution: India (WB).
221. *E. blutanensis* (Bielawski)
Distribution: Bhutan.
222. *E. brivoioi* (Bielawski and Fursch)
Distribution: Burma.
223. *E. cherranpunjiensis* (Kapur)
Distribution: India.
224. *E. congener* (Gorham)
Distribution: India (SK, WB) and Burma.
225. *E. convextata* (Singh and Singh)
Distribution: India.
226. *E. crecentomaculata* (Singh and Singh)
Distribution: India.
227. *E. decemmaculata* (Redtenbacher)
Distribution: India (J&K), Sri Lanka, Burma and China.
228. *E. delessertii* (Guerin-Meneville)
Distribution: India and Sri Lanka.
229. *E. deyrollii* (Crotch)
Distribution: India (SK, WB) and Burma.
230. *E. dorotae* (Bielawski)
Distribution: Bhutan.
231. *E. dumerili* (Mulsant)
Distribution: India, Sri Lanka, Nepal, Burma, China and Vietnam.

232. *E. elvina* (Mulsant)

Distribution: India and Nepal.

233. *E. endomycina* (Gorham)

Distribution: India.

234. *E. erinacea* (Sicard)

Distribution: South India.

235. *E. fasciolata* (Crotch)

Distribution: India.

236. *E. flavicollis* (Thunberg)

Distribution: India, Sri Lanka, Burma, Thailand, China, Vietnam, Philippines, Indonesia and Taiwan.

237. *E. gibbera* (Crotch)

Distribution: India (SK) and Nepal.

238. *E. gokteika* (Kapur)

Distribution: Burma and Vietnam.

239. *E. gorkhana* (Miyatake)

Distribution: Nepal.

240. *E. grayi* (Mulsant)

Distribution: India (AS, UP, WB), Nepal, China and Indonesia.

241. *E. hendecaspilota* (Mader)

Distribution: India, Sri Lanka, Burma, Nepal, Thailand, Vietnam, China, Taiwan and Philippines.

242. *E. hingstoni* (Kapur)

Distribution: Burma, Taiwan and Indonesia.

243. *E. hopeiana* (Miyatake)

Distribution: Nepal.

244. *E. ipmicticollis* (Sicard)

Distribution: India.

245. *E. incauta* (Mulsant)

Distribution: Burma, Taiwan and Indonesia.

246. *E. intermixta* (Dieke)

Distribution: India.

247. *E. laesicollis* (Mulsant)

Distribution: India (SK, WB) and Nepal.

248. *E. laosana* (Bielawski)

Distribution: India and Vietnam.

249. *E. laculosa* (Sicard)
Distribution: Indian(TN: Nilgiris).
250. *E. macularis* (Mulsant)
Distribution: India, Nepal and China.
251. *E. maculivestis* (Mulsant)
Distribution: India, Nepal and China.
252. *E. marginicollis* (Hope)
Distribution: India (SK, WB), Nepal, Bhutan and Burma.
253. *E. mausmaiensis* (Jadwiszczak)
Distribution: India (AS).
254. *E. maxima* (Weise)
Distribution: India (AS), China, Taiwan and Vietnam.
255. *E. militaris* (Dieke)
Distribution: India and Nepal.
256. *E. monsuna* (Bielawski)
Distribution: Bhutan.
257. *E. moorii* (Crotch)
Distribution: India: WB.
258. *E. mystica*(Mulsant)
Distribution: India (Northern region), Nepal, Bhutan and Burma.
259. *E. nepalensis* (Kapur)
Distribution: Nepal.
260. *E. nevilli* (Dohrn)
Distribution: India (Andamans).
261. *E. nielamuensis* (Pang and Mao)
Distribution: Nepal. China.
262. *E. nilgirica* (Weise)
Distribution: South India.
263. *E. octamaculata* (Thurnberg)
Distribution: Burma.
264. *E. pembertoni* (Crotch)
Distribution: India and Bhutan.
265. *E. probsti* (Jadwiszczak and Putz)
Distribution: India (Karnataka).
266. *E. ryszardi* (Jadwiszczak and Wegrzynowicz)
Distribution: Bhutan.

267. *E. schawallerri* (Canepari)
    Distribution: Nepal.
268. *E. septemnotata* (Bielawski)
    Distribution: Bhutan.
269. *E. septemocellata* (Singh and Singh)
    Distribution: India (HP).
270. *E. sexpunctata* (Bielawski)
    Distribution: Bhutan.
271. *E. sexpustulata* (Bielawski)
    Distribution: Bhutan.
272. *E. shilliensis* (Singh and Singh)
    Distribution: India (HP).
273. *E. simia* (Sicard)
    Distribution: India (TN: Anamalais).
274. *E. subclathrata* (Sicard)
    Distribution: India (TN: Nilgiri Hills).
275. *E. sureilica* (Kapur)
    Distribution: India and Burma.
276. *E. suspiciosa* (Weise)
    Distribution: India (SK, WB).
277. *E. undecimspilota* (Hope)
    Distribution: India and Nepal.
278. *E. wallichii* (Hope)
    Distribution: Nepal.
279. *E. zeylanica* (Crotch)
    Distribution: Sri Lanka.
280. *Henosepilachana anita* (Kapur)
    Distribution: South India.
281. *H. balajuensis* (Jadwiszczak and Putz)
    Distribution: Nepal.
282. *H. bielawskii* (Fursch)
    Distribution: South India.
283. *H. boisduvali* (Mulsant)
    Distribution: India, China, Taiwan, Vietnam, Philippines, Indonesia, New Guinea, Samoa Islands, New Hebrides, Fiji Islands and Australia.

284. *H. circellaris* (Korschefsky)

Distribution: India (Karnataka).

285. *H. circularis* (Korchefsky)

Distribution: India (Karnataka).

286. *H. dodecastigma* (Wiedemann)

Distribution: India and Bangladesh.

287. *H. dubiosa* (Dieke)

Distribution: India and Indonesia (Borneo).

288. *H. elaterii* (P.Rossi)

Distribution: Pakistan, Afghanistan, South Europe, Turkey, North Africa, Sudan, Senegal, Tschad, Niger, Volta, Iran and Iraq.

289. *H. gangetica* (Weise)

Distribution: India (PY, TN) and Vietnam.

290. *H. implitica* (Mulsant)

Distribution: India.

291. *H. indica* (Mulsant)

Distribution: India (Northeastern region), Myanmar, China, Malaysia and Indonesia (Java).

292. *H. karenensis* (Kapur)

Distribution: Burma.

293. *H. kaszabi* (Bielawski)

Distribution: India (Andaman), Burma, Philippines, China and Vietnam.

294. *H. kathmandueensis* (Miyatake)

Distribution: Nepal.

295. *H. khasiensis* (Weise)

Distribution: India (AS).

296. *H. manipurensis* (Kapur)

Distribution: India (Manipur).

297. *H. nana* (Kapur)

Distribution: South India.

298. *H. narayana* (Lal and Kanakavalli)

Distribution: India (KA, MH).

299. *H. ocellata* (Redtenbacher)

Distribution: India: Northern region. Nepal. China.

300. *H. ornate* (Dieke)

Distribution: India.

301. *H. perplexa* (Dieke)

Distribution: India and Thailand.

302. *H. pusillanima* (Mulsant)

Distribution: India, Nepal, China, Taiwan, Thailand, Vietnam, Philippines and Indonesia.

303. *H. pytho* (Mulsant)

Distribution: Burma, Indonesia and Philippines.

304. *H. quarta* (Dieke)

Distribution: India.

305. *H. quinta* (Dieke)

Distribution: Sri Lanka.

306. *H. septima* (Dieke)

Distribution: India, Sri Lanka and Vietnam.

307. *H. sikkimica* (Kapur)

Distribution: India (SK).

308. *H. vigintioctomaculata* (Motschulsky)

Distribution: India, Nepal, China, Taiwan, Vietnam, Korea, Japan and Russian Far East.

309. *H. vigitioctopunctata* (Fabricus)

Distribution: India, Sri Lanka, Pakistan, Nepal, Bhutan, Burma, Thailand, Vietnam, Philippines, Indonesia, Japan, China, New Guinea, Fiji, Solomon Islands and Australia.

310. *H. zebra* (Sicard)

Distribution: South India.

311. *Macrolasia arcula* (Weise)

Distribution: India (PY).

Subfamily: Scymninae

Tribe: Aspidimerini

312. *Aspidimerus birmanicus* (Gorham)

Distribution: Myanmar and Thailand.

313. *A. nigrovittatus* (Motschulsky)

Distribution: Sri Lanka.

314. *A. rufircus* (Gorham)

Distribution: Myanmar.

315. *A. spencei* (Mulsant)

Distribution: India (AS, ME: Khasia Hills) and Myanmar.

316. *Cryptogonus ariasi* (Mulsant)
Distribution: India (SK, UP), Myanmar and Nepal.
317. *C. bilineatus* (Kapur)
Distribution: India (TN).
318. *C. bimaculatus* (Kapur)
Distribution: India (AS, TN), Bhutan, Myanmar, Nepal and Thailand.
319. *C. bryanti* (Kapur)
Distribution: Sri Lanka.
320. *C. complexus* (Kapur)
Distribution: India (AS, SK), Bhutan, Myanmar and Nepal.
321. *C. downingi* (Kapur)
Distribution: Myanmar.
322. *C. fulvoterminatus* (Boheman)
Distribution: Myanmar, Thailand, Indonesia and Malaysia.
323. *C. himalyaensis* (Kapur)
Distribution: India (SK), Bhutan and Myanmar.
324. *C. hingstoni* (Kapur)
Distribution: India (SK).
325. *C. kapuri* (Ghorpade)
Distribution: India (KA).
326. *C. laetus* (Weise)
Distribution: Myanmar.
327. *C. lepidus* (Weise)
Distribution: India (AS), Myanmar and Thailand.
328. *C. loebli* (Canepari)
Distribution: India (UP: Garhwal and Kumaon).
329. *C. nepalensis* (Bielawski)
Distribution: Nepal.
330. *C. nepalensis bhutanensis* (Bielawski)
Distribution: Bhutan
331. *C. nitidus* (Kapur)
Distribution: Myanmar and Nepal.
332. *C. orbiculus* (Gyllenhal)
Distribution: India (AS, ME), Bhutan, Nepal, Sri Lanka, China, Taiwan, Japan and Micronesia.

333. *C. postmedialis* (Kapur)

Distribution: India (AS, SK, UP, WB), Bhutan, Nepal and Myanmar.

334. *C. quadriguttatus* (Weise)

Distribution: India (SK, AS, WB, UP, GA) and Bhutan.

335. *C. trioblitus* (Gorham)

Distribution: India (AS, UP), Nepal, Myanmar and Thailand.

336. *Pseudaspidimerus flaviceps* (Walker)

Distribution: India (TN) and Sri Lanka.

337. *P. infuscatus* (Poorani)

Distribution: India (TN).

338. *P. lambai* (Kapur)

Distribution: India (AN).

339. *P. mauliki* (Kapur)

Distribution: India (AS, ME), Bangladesh and Thailand.

340. *P. triotatus* (Thunberg)

Distribution: India (AS, KL, KA, AN, WB, GA), Sri Lanka, Bangladesh, Myanmar.

341. *P. uttami* (Kapur)

Distribution: India (TN, KA) and Sri Lanka.

Tribe: Hyperaspidini

342. *Hyperaspis maindroni* (Sicard)

Distribution: India (AS, KA, TN) and Pakistan.

343. *H. marginaloides* (Canepari)

Distribution: Nepal.

Tribe: Ortaliini

344. *Amida vietnamica* (Hoang)

Distribution: India (Northeast), China and Vietnam.

345. *Ortalia andrewesi* (Sicard)

Distribution: Myanmar and Thailand.

346. *O. chlorops* (Gorham)

Distribution: India (TN: Anamalais).

347. *O. discoidea* (Weise)

Distribution: India (BR).

348. *O. horni* (Weise)

Distribution: India (TN, AS, ME, WB), Sri Lanka, Myanmar, Thailand and China.

349. *O. kandyana* (Bielawski)

Distribution: Sri Lanka.

350. *O. maeklini* (Mulsant)

Distribution: India (UP), Myanmar, Thailand and Indonesia.

351. *O. minuta* (Weise)

Distribution: Sri Lanka.

352. *O. pectoralis* (Weise)

Distribution: India (ME).

353. *O. pusilla* (Weise)

Distribution: India (UP), Indonesia and Philippines.

354. *O. quadripunctata* (Gorham)

Distribution: India (KA, TN).

Tribe: Scymnini

355. *Axinoscymnus puttarudriahi* (Kapur and Minshi)

Distribution: India (KA, TN, KL, LDI) and Sri Lanka.

356. *Clitostethus fumatus* (Sicard)

Distribution: Sri Lanka.

357. *C. lividipes* (Sicard)

Distribution: Myanmar.

358. *C. luteoniger* (Canepari)

Distribution: Nepal.

359. *Cryptolaemus montrouzieri* (Mulsant)

Distribution: Native of Indo-Australian region, Introduced into India, the USA, UK and many other countries.

Genus Diomus (Mulsant) has not been reported from the Indian region.

360. *Horniolus dispar* (Weise)

Distribution: Sri Lanka.

361. *H. guimeti* (Mulsant)

Distribution: India (KA, TN) and Bangladesh.

362. *H. hisamatsui* (Miyatake)

Distribution: Nepal and Hong Kong.

363. *H. nigreipes* (Miyatake)
Distribution: India (TN).
364. *Nephus bistillatus* (Mulsant)
Distribution: India.
365. *N. lentiformis* (Booth)
Distribution: India (TN).
366. *N. patruus* (Weise)
Distribution: Sri Lanka.
367. *N. regularis* (Sicard)
Distribution: India (AP, AS, KA, TN), Pakistan and China.
368. *N. severini* (Weise)
Distribution: India.
369. *N. similincludens* (Canepari)
Distribution: Nepal.
370. *N. tagiapatus* (Kamiya)
Distribution: India (WB, AN), Thailand, Malaysia and Japan.
371. *Psedoscymnus dwipakalpa* (Ghorpade)
Distribution: India (AP, KA, TN).
372. *P. funerarius* (Canepari)
Distribution: Nepal.
373. *P. luteus* (Sicard)
Distribution: India (KL) and Celebes.
374. *P. murreensis* (Ahmad)
Distribution: Pakistan.
375. *P. nepalicus* (Miyatake)
Distribution: Nepal.
376. *P. ocelliferus* (Canepari)
Distribution: Nepal.
377. *P. pallidicollis* (Mulsant)
Distribution: India.
378. *P. simmondsi* (Chapin and Ahmad)
Distribution: Pakistan and India.

Subgenus *Neopullus*

379. *Scymnus* (*Neopullus*) *fuscatus* (Bohema)
Distribution: India (AS, KA, AN, WB), Sri Lanka, Pakistan, Nepal, China, Taiwan, Philippines, Japan and Thailand.

380. *S.* (*Neopullus*) *hoffmanni* (Weise)
    Distribution: India, China and Japan.
381. *S.* (*Neopullus*) *loebli* (Canepari)
    Distribution: India (UP).

Subgenus *Pullus*

382. *Scymnus* (*Pullus*) *assamensis* (Canepari)
    Distribution: India (AS: Manas).
383. *S.* (*Pullus*)*bengalensis* (Canrpari)
    Distribution: India (WB).
384. *S.* (*Pullus*)*besucheti* (Canepari)
    Distribution: India (WB) and Nepal.
385. *S.* (*Pullus*) *bourdilloni* (Kapur)
    Distribution: India (SK) and Nepal.
386. *S.* (*Pullus*) *castaneus* (Sicard)
    Distribution: India (TN, AP, KA, KL) and Bangladesh.
387. *S.* (*Pullus*) *ceylonicus* (Motschulsky)
    Distribution: India (KA) and Sri Lanka.
388. *S.* (*Pullus*) *coccivora* (Ayyar)
    Distribution: India (AP, KA, TN, GA, UP), Pakistan, Bangladesh, Sri Lanka and Malaysia.
389. *S.* (*Pullus*) *exilis* (Canepari)
    Distribution: Nepal.
390. *S.* (*Pullus*) *facetus* (Canepari)
    Distribution: India (UP, J&K)
391. *S.* (*Pullus*) *fruticis* (Canepari)
    Distribution: Nepal.
392. *S.* (*Pullus*) *giganteus* (Kamiya)
    Distribution: India and Japan.
393. *S.* (*Pullus*) *godavariensis* (Miyatake)
    Distribution: Nepal.
394. *S.* (*Pullus*) *gracilis* (Motschulsky)
    Distribution: India.
395. *S.* (*Pullus*) *gratiosus* (Weise)
    Distribution: Sri Lanka.
396. *S.* (*Pullus*) *harejoides* (Sicard)
    Distribution: India.

397. *S. (Pullus) hepaticus* (Canepari)
 Distribution: Nepal.
398. *S. (Pullus) hilaris* (Motschulsky)
 Distribution: India (SK), Myanmar and Sri Lanka.
399. *S. (Pullus) hingstoni* (Kapur)
 Distribution: India (SK, ME).
400. *S. (Pullus) impexus* (Mulsant)
 Distribution: India (UP), Europe and North Africa.
401. *S. (Pullus) janetscheki* (Bielawski)
 Distribution: Nepal.
402. *S. (Pullus) kawamurai* (Ohta)
 Distribution: India (UP, J&K) and Japan.
403. *S. (Pullus) kosianus* (Canepari)
 Distribution: Nepal.
404. *S. (Pullus) latemaculatus* (Motschulsky)
 Distribution: India (KA, TN, LDI, GA), Bangladesh, Sri Lanka, Thailand and Taiwan.
405. *S. (Pullus) lucicolus* (Canepari)
 Distribution: Nepal.
406. *S. (Pullus) martensi* (Canepari)
 Distribution: Nepal.
407. *S. (Pullus) mehalayae* (Canepari)
 Distribution: India (ME: Khasi Hills).
408. *S. (Pullus) mellinus* (Canepari)
 Distribution: Nepal.
409. *S. (Pullus) musculus* (Weise)
 Distribution: Sri Lanka.
410. *S. (Pullus) nepalensis* (Bielawski)
 Distribution: India (WB) and Nepal.
411. *S. (Pullus) nymphaens* (Kapur and Mulsant)
 Distribution: India (WB)
412. *S. (Pullus) o-nigrum* (Mulsant)
 Distribution: India.
413. *S. (Pullus) posticalis* (Sicard)
 Distribution: India (HP, AS), Nepal, China, Japan and Myanmar.

414. *S.* (*Pullus*) *pyrochelus* (Mulsant)
Distribution: India (AN, WB), Nepal and Bangladesh.
415. *S.* (*Pullus*) *quadrillum* (Motschulsky)
Distribution: Nepal
416. *S.* (*Pullus*) *rufoniger* (Canepari)
Distribution: Nepal
417. *S.* (*Pullus*) *saciformis* (Motschulsky)
Distribution: India and Sri Lanka.
418. *S.* (*Pullus*) *sapporensis* (Ohta)
Distribution: India (Northeastern region), Japan and China.
419. *S.* (*Pullus*) *schawalleri* (Canepari)
Distribution: Nepal.
420. *S.* (*Pullus*) *sodalist* (Weise)
Distribution: India (UP: Garhwal), Japan and Taiwan.
421. *S.* (*Pullus*) *suturaloides* (Canepari)
Distribution: Nepal.
422. *S.* (*Pullus*) *testsacecollis* (Kapur)
Distribution: India (SK.
423. *S.* (*Pullus*) *victoris* (Crotch)
Distribution: India (Nicobar Islands), Himalayas, Sri Lanka, Philippines and Nepal.
424. *S.* (*Pullus*) *victoris unimaculata* (Korschefsky)
Distribution: India.
425. *S.* (*Pullus*) *uninotata* (Gorham)
Distribution: India (KA).
426. *S.* (*Pullus*) *weigeli* (Canepari)
Distribution: Nepal.
427. *S.* (*Pullus*) *xerampelinus* (Mulsant)
Distribution: India (KA, TN), Bangladesh.
428. *S.* (*Pullus*) *zonatus* (Sicard)
Distribution: India.

Subgenus *Orthoscymnus*

429. *Scymnus* (*Orthoscymnus*) *rhododendri* (Canepari)
Distribution: Nepal.
430. *S.* (*Orthoscymnus*) *smetanai* (Canepari)
Distribution: Nepal.

Subgenus *Scymnus*

431. *S.* (*Scymnus*) *albopilis* (Motschulsky)
 Distribution: Sri Lanka.
432. *S.* (*Scymnus*) *andrewesi* (Sicard)
 Distribution: India (KA, TN).
433. *S.* (*Scymnus*) *apiciflavus* (Motschulsky)
 Distribution: India, Sri Lanka, Thailand and China.
434. *S.* (*Scymnus*) *indicus* (Weise)
 Distribution: India (Eastern India, KA).
435. *S.* (*Scymnus*) *lepidulus* (Motschulsky)
 Distribution: India (BR) and Sri Lanka.
436. *S.* (*Scymnus*) *levaillantia* (Mulsant)
 Distribution: India (UP: Garhwal), Europe and Africa.
437. *S.* (*Scymnus*) *nubilus* (Mulsant)
 Distribution: India (AS, KA, LDI, MH, OR, TN, WB, GA), Pakistan, Bangladesh, Sri Lanka, Nepal, Myanmar, China and Asia Minor.
438. *S.* (*Scymnus*) *sufflavus* (Canepari)
 Distribution: Nepal.
439. *S.* (*Scymnus*) *tristigmaticus* (Gorham)
 Distribution: India (ME: Khasi Hills), Nepal and Myanmar.
440. *S.* (*Scymnus*) *trisulicus* (Canepari)
 Distribution: Nepal.
441. *S.* (*Scymnus*) *venalis* (Mulsant)
 Distribution: India.

Subgenus not known

442. *S. asciuianus* (Gorham)
 Distribution: Myanmar.
443. *S. seriatus* (Weise)
 Distribution: E. India ("Bengal").

Tribe: Stethorini

444. *Stethorus gilvifrons* (Mulsant)
 Distribution: India (AN, KA, KL, HP), Pakistan, Italy and Cyprus.
445. *S. guangxiensis* (Pang and Mao)
 Distribution: India (KA) and China.

446. *S. indira* (Kapur)
Distribution: India (WB, ME), China and Thailand.
447. *S. keralicus* (Kapur)
Distribution: India (KL, TN, KA).
448. *S. parcepunctatus* (Kapur)
Distribution: India (KA).
449. *S. puaperculus* (Weise)
Distribution: India (KA, TN, BR, DL, MH), Pakistan, Thailand and Malaysia.
450. *S. rani* (Kapur)
Distribution: India (UP), Thailand and China.
451. *S. tetranychi* (Kapur)
Distribution: Bangladesh and Thailand.

Subfamily: Sticholotidinae (=Sticholotinae)

Tribe: Plotinini (Miyatake)

452. *Buprestodera inornata* (Miyatake)
Distribution: India (TN).
453. *B. mimetica* (Sicard)
Distribution: India (KA, TN).
454. *Paraplotina flavomaculata* (Miyatake)
Distribution: India (TN).
455. *Protoplotina nigrosuturalis* (Poorani)
Distribution: India (UP).
456. *P. vietnamica* (Miyatake)
Distribution: India (ME).

Tribe: Serangiini

457. *Catana chapini* (Kapur)
Distribution: India (UP).
458. *Microserangium brunneonigrum* (Poorani)
Distribution: India (TN).
459. *M. laterale* (Motschulsky)
Distribution: Sri Lanka.
460. *Serangium montazerii* (Fursch)
Distribution: India (UP), Pakistan, Iran, Syria, Introduced into Israel and Europe.

461. *S. parcesetosum* (Sicard)

Distribution: India (MH, TN, MP, UP, LDI), Pakistan, Thailand, Introduced into Turkey and the USA.

462. *S. serratum* (Poorani)

Distribution: India (KA, TN).

Tribe: Shirozuellini (=Ghaniini)

463. *Ghanius karachiensis* (Ahmad)

Distribution: Pakistan.

464. *G. schawalleri* (Canepari)

Distribution: Nepal. India (a new species found to occur in south India).

465. *Promecopharus andrewesi* (Sicard)

Distribution: India.

Tribe: Sticholotidini (=Pharini)

466. *Jauravia albidula* (Motschulsky)

Distribution: Sri Lanka.

467. *J. assamensis* (Kapur)

Distribution: India (AS).

468. *J. dorsalis* (Weise)

Distribution: India (TN, PY, KA), and Myanmar.

469. *J. hanifi* (Afroze and Shujauddin)

Distribution: India (UP).

470. *J. indica* (Kapur)

Distribution: India (TN).

471. *J. kanarensis* (Kapur)

Distribution: India (KA).

472. *J. Kapuri* (Iablokoff- Khnzorian)

Distribution: Sri Lanka.

473. *J. limbata* (Motschulsky)

Distribution: India (AS, KA, KL, GA), Sri Lanka, Nepal, Taiwan and Thailand.

474. *J. opaca* (Weise)

Distribution: India (KA, MH) and Sri Lanka.

475. *J. pallidula* (Motschulsky)

Distribution: India (TN, KA, WB) and Sri Lanka.

476. *J. pilosulla* (Weise)
Distribution: India (TN) and Sri Lanka.
477. *J. pubescens* (Fabricus)
Distribution: India (TN, OR, KA) and Sri Lanka.
478. *J. quadrinotata* (Kapur)
Distribution: India (AS, WB, ME), Bhutan and Nepal.
479. *J. simplex* (Walker)
Distribution: India (KA, MH) and Sri Lanka.
480. *J. soror* (Weise)
Distribution: India (TN, KA, KL) and Sri Lanka.
481. *Parajauravia testivestis* (Mulsant)
Distribution: India.
482. *Pharoscymnus flexibilis* (Mulsant)
Distribution: India (BR, J&K, KA, AP) and Pakistan.
483. *P. flexibilis kashmirensis* (Kapur)
Distribution: India (J&K, UP) and Pakistan.
484. *P. horni* (Weise)
Distribution: Pakistan and Thailand.
485. *P. simmondsi* (Ahmad)
Distribution: Pakistan and Thailand.
486. *P. suturalis* (Sicard)
Distribution: Sri Lanka and Myanmar.
487. *Sticholotis amator* (Kapur)
Distribution: Nepal.
488. *S. besucheti* (Canepari)
Distribution: India (AS: Manas); ME: Garo Hills).
489. *S. bilineata* (Weise)
Distribution: India (SK).
490. *S. binotata* (Gorham)
Distribution: India (KA).
491. *S. cribellata* (Sicard)
Distribution: India (KA).
492. *S. decempunctata* (Weise)
Distribution: India and Nepal.
493. *S. decora* (Weise)
Distribution: India (TN: Nilgiri Hills).

494. *S. disoidea* (Gorham)
Distribution: Myanmar.
495. *S. dohrni* (Weise)
Distribution: Myanmar.
496. *S. duodecimmaculata* (Canepari)
Distribution: Nepal.
497. *S. dudecimpunctata* (Weise)
Distribution: India (SK, WB), Himalayas and Nepal.
498. *S. exsanguis* (Sicard)
Distribution: India.
499. *S. ferruginea* (Gorham)
Distribution: India (KA).
500. *S. honesta* (Weise)
Distribution: India and Thailand.
501. *S. horni* (Weise)
Distribution: Sri Lanka.
502. *S. kumatai* (Miyatake)
Distribution: Nepal.
503. *S. marginalis* (Kapur)
Distribution: India (J&K, HP), Pakistan and Nepal.
504. *S. minuta* (thurnberg)
Distribution: "India Orientali".
505. *S. nepalensis* (Miyatake)
Distribution: Nepal.
506. *S. nilgriensis* (Weise)
Distribution: India (TN: Nilgiris).
507. *S. obscurella* (Weise)
Distribution: India (TN: Nilgiris).
508. *S. obcurocincta* (Sicard)
Distribution: India.
509. *S. quadrisignata* (Weise)
Distribution: India and Vietnam.
510. *S. quadrisignata rugicollis* (Korschefsky)
Distribution: India (KA).
511. *S. rufoplagiata* (Motschulsky)
Distribution: Sri Lanka.

512. *S. sanguinolenta* (Motschulsky)
Distribution: India.
513. *S. sanguinosa* (Motschulsky)
Distribution: India.
514. *S. sexpunctata* (Sicard)
Distribution: India.
515. *S. substriata* (Crotch)
Distribution: India (TN), Japan and China.
516. *S. transversa* (Motschulsky)
Distribution: India.
517. *S. tredecimmaculata* (Weise)
Distribution: India (TN: Nilgiri Hills).
518. *Stictobura gibbula* (Weise)
Distribution: India (TN).
519. *S. melannaria* (Weise)
Distribution: India (PY).
520. *S. pallideguttata* (Mulsant)
Distribution: India (TN, KL).
521. *S. rubroguttata* (Sicard)
Distribution: India (TN: Nilgiris).
522. *S. semipolita* (Sicard)
Distribution: India (TN, KA, KL).
523. *Synonychimorpha chittagongi* (Vazirani)
Distribution: India (KA, TN) and Nepal.
524. *S. immaculate* (*Poorani*)
Distribution: India (KA, KL).

Tribe: Sukunahikonini

525. *Scotoscymnus popei* (Vazirani)
Distribution: India (GJ).

Chapter 6

Nature of Ladybird Beetle

Reflex Bleeding

The bitter taste and toxic properties of many ladybird beetles can be attributed to a variety of defensive alkaloids adaline, coccinelline, exochomine, hippodamine, *etc.*) (Daloze *et al.,* 1995). About 50 different alkaloids have been identified in coccinellids (Laurent *et al.,* 2005). The alkaloids are produced in ladybird haemolymph and are distributed by the haemolymph throughout the insect's body. When disturbed, ladybirds secrete droplets of haemolymph through tibio -femoral articulations. The released fluid is called 'reflex blood' and the corresponding defense mechanism 'reflex bleeding'. Reflex bleeding is commonly used by adults of many ladybird species and by larvae and/or pupae of some species, but in the latter cases, the fluid is usually released from pores in the dorsal body surface (Holloway *et al.,* 1991, Daloze *et al.,* 1995). The pupae of some Epilachninae are known to exude the droplets of defensive fluid by specialized, glandular hairs (Schroeder *et al.,* 1998). In adult ladybirds, reflex bleeding is often associated with thanatosis (Daloze *et al.,* 1995, Ceryngier and Hodek 1996).

The high toxicity of *C. septempunctata* was associated with severe pathological changes in the blue tits' livers (Marples *et al.,* 1989), and this was probably caused by coccinelline, the main *C. Septempunctata* alkaloid. In contrast to coccinelline, the major alkaloid compound of *Adalia* species, adaline, appears rather benign to young tits. Reflex blood also contains methylalkylpyrazines, which are involved in ladybird defense. In contrast to alkaloids, pyrazines are volatile, so they can be olfactorily detected by predators. They are responsible for the odour that, in addition to aposematic colouration, acts as a signal in highlighting the unprofitability of ladybird as food. It was found that pyrazines tend to be absent from cryptically coloured Coccinellidae, but are often present in aposematic species (Moore *et al.,*

1990, Daloze *et al.,* 1995).Reflex bleeding may also act as a mechanical defense against some invertebrate predators. The fluidcoagulates quickly on exposure to air and may stick toa predator's legs, antennae and mouthparts (Eisner *et al.,* 1986).

The immature stages (eggs, larvae, and pupae) also contain the toxins that their adults have, and in this characteristic, they resemble rove beetles (Staphylinidae) of the genus *Paederus* (Farnk and Kanamitsu 1987) although the toxins are totally different. Toxins are said to be produced by dorsal glands in the larvae.

Behaviour and Communication

As cold-blooded species, ladybird beetles mainly are diurnal, utilizing as much sunlight as possible for feeding and mating. For this reason and for protection, ladybird beetle are form aggregate communities of up to 15 individuals for the overwintering period. They find protected areas of dense foliage, giving of chemical cues to attract other individuals to the area.

Range of Coloring

Although ladybirds are often thought of as being red, they also appear in a multitude of other colors depending on their species. These colors include both orange and yellow. To a lesser degree, they include black, grey, pink and blue *etc.,*

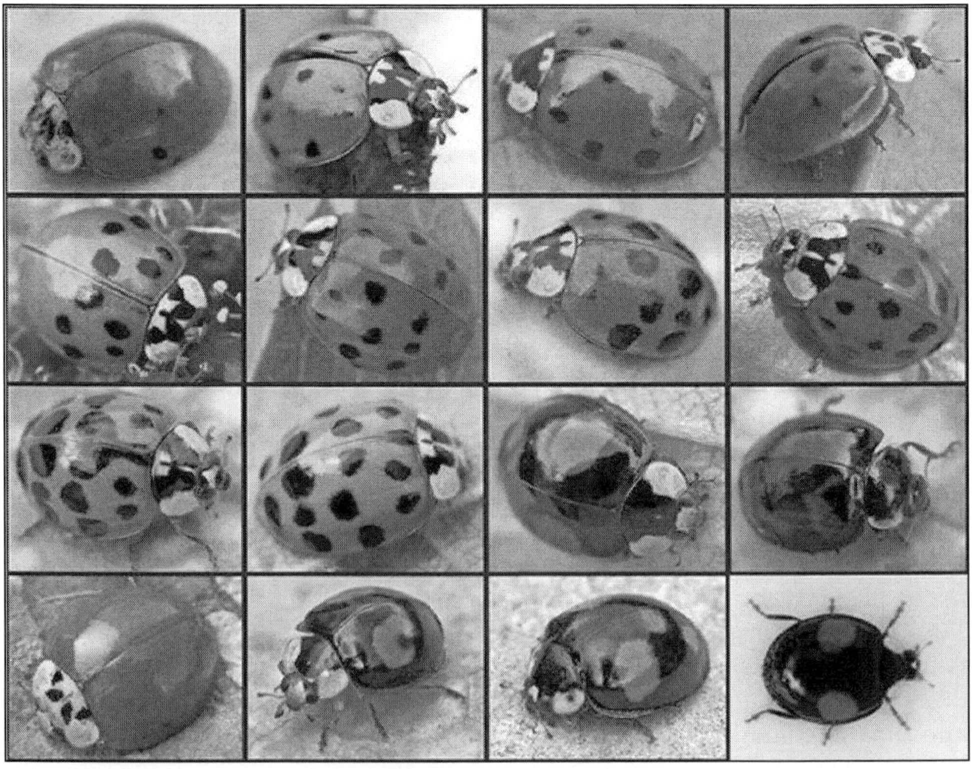

Figure 9: Range of Coloring of Ladybird.

Ladybird beetle an also feature flecks on the upper parts of their wings, whether in pink, red, black, yellow or white. Many ladybird beetles just have two flecks, while many others have as many as 13. Ladybird beetles occasionally can be totally fleck-free. Whatever the exact colour scheme and design, ladybird beetles are physically conspicuous creatures.

Ladybird Beetle Spots Warn Potential Predators

The spots and bright colors are meant to warn would be attackers that this beetle tastes terrible. They have also been known to play dead, giving them a two pronged defense system in a world of eat or be eaten. Because of this, they aren't often preyed upon, however several insects, such as Assassin bugs, Stink bugs and spiders may eat ladybird beetles.

Cannibalistic Nature

Cannibalism of eggs, larvae and pupae is common, especially when prey is scarce. Although the occurrence of egg cannibalism and predation most often focuses on larvae, adults also come into contact with coccinellid eggs. However, less is known about egg cannibalism and predation by adult coccinellidae, especially native egg predation by exotic adults. In addition, even less is known about egg cannibalism and predation by adult male and female coccinellids although it could be expected that males would feed less.

Figure 10: Coccinellid Larva Feed on their Pupa.

Chapter 7

Semiochemicals and Ladybird Beetles

Semiochemicals play key roles in the conspecific, heterospecific, and prey interactions of ladybirds. Their main role is related to attack, food, sex, and defense (Hemptinne and Dixon, 2000). Pattanayak *et al.* (2014, 2015) employed headspace solid-phase microextraction to determine sex-specific volatile hydrocarbon profile of five ladybirds from India and found methyl-branched saturated and unsaturated hydrocarbons. They listed 26 alkane-based compounds, which varied in composition in ladybirds, including *Menochilus sexmaculatus* (Fabr.), *C. septempunctata*, *Coccinella transversalis* Fab. and *Propylea dissecta* (Mulsant). In *Anegleis cardoni* (Weise) the two sexes differ in major chemicals, but both the sexes have similar hydrocarbon (HC) profiles and 9- and 7-methyl tricosane predominate the cuticular HC profile in of *A. bipunctata* (Hemptinne *et al.*, 1998). The HCs are key volatiles and are found in the footprints (Magro *et al.*, 2010) and cuticular lipids (Durieux *et al.*, 2012).

Kairomones Driven Foraging

Volatile infochemicals generated by prey, their host plants, or from the interaction between host plant and herbivore may be used by predators in prey location. Ladybirds are more attracted toward the plants where the aphids are disturbed than those where aphids are undisturbed or toward healthy plants (Ninkovic *et al.*, 2001). Plants and aphids release volatiles on disturbance which are perceived by the ladybirds. Plants induce two classes of volatile compounds (IVCs) (Dudareva *et al.*, 2006). Damaged leaves release class I IVCs mainly consisting of C6 green volatile compounds, while class II IVCs (mainly terpenes) are synthesized and released after a few hours of plant injury (Holopainen, 2004). Ladybirds are attracted toward chemicals, like limonene and β-caryophyllene, for prey location and oviposition (Francis *et al.,* 2005; Alhmedi *et al.,* 2010b). Limonene is a potential kairomone for *H. axyridis* and females are found more in limonene zones rather than in control zones.

This explains why they hover more in citrus orchards that are a rich source of limonene (Michaud, 2002). When attacked, aphids release a small droplet from their abdominal cornicles containing an alarm pheromone, identified as the sesquiterpene or (E)-β-farnesene (Francis *et al.,* 2005). Conspecific aphids trigger their escape behaviors after perceiving it. This kairomone attracts adult ladybirds, who display electrophysiological responses toward a group of semiochemicals including both host and herbivore odors. This additively mitigates the attraction of ladybirds toward aphid colonies (Joachim and Weisser, 2013). Nonetheless, feeding by ladybirds is largely driven by this kairomone.

Oviposition Deterring Pheromones

Oviposition-deterring pheromones (ODP) are chemicals that are released from the invaders that deter the mother ladybird to oviposit in such sites (Ruzicka, 2002; Michaud and Jyoti, 2007; Ruzicka and Zemek, 2008). It is a part of maternal strategy in a bid to protect her offspring from the enemies. Females search for the most suitable oviposition sites that may favor their progeny (Kindlmann and Dixon, 1993). This site selection depends on abiotic conditions, nutritional requirements of offspring and resource availability (Dixon, 2000), absence of predators, and plant architecture (Seagraves, 2009).

ODPs of both conspecifics and heterospecifics invaders are indicators of age and future of prey colony. These are persistent in both larval (Mishra *et al.,* 2012) and adult footprints (Mishra and Omkar, 2006; Mishra *et al.,* 2012) and are released from anal disk on the tenth abdominal segment of larvae (Laubertie *et al.,* 2006). These are a mixture of HCs dominated by n-pentacosane producing a persistent signal (Hemptinne *et al.,* 2001). Additionally, n-tricosane and n-heptacosane are also important footprint semiochemicals in ladybirds (Nakashima *et al.,* 2006). n-Z-12-pentacosene is a principal compound in *M. sexmaculatus* (Pattanayak *et al.,* 2014) and is considered as an important ODP (Klewer *et al.,* 2007). These footprints also act as foraging-deterrent pheromones and reduce feeding activity and delay development in ladybirds (Kumar *et al.,* 2014).

The effects of ODP are largely density and stage-dependent (Doumbia *et al.,* 1998; Mishra and Omkar, 2006). However, the possibility of the sex-dependent sensitivity toward these chemicals is not totally ignored (Hodek and Michaud, 2008). Aged and experienced mothers may modify the response to ODP (Frechette *et al.,* 2004). Female ladybirds usually deter from ovipositing on plants contaminated with conspecific larval tracks but not with heterospecific tracks (Yasuda *et al.,* 2000).

Chapter 8

Prey Specificity of Coccinellids

In fact, as soon as ladybird beetles hatch, they begin to feast. Adult females lay eggs hundreds of them in aphid colonies, and when they hatch, the larvae immediately start feeding. "Once hatched, the larvae eat about 350 to 400 aphids in the two weeks it takes them to become fully grown. But they don't just eat aphids, they also eat fruit flies, thrips, mites and other plant damaging insects.

However, different species prefer different foods. While many prey on garden pests, like Mexican beetle and the squash beetle feed on the plant leaves mentioned in their names, making those particular species unwelcome guests in some fields.

Rodalia cardinalis against *Icerya purchasi*

The vedalia beetla, *Rodalia cardinalis*, was introduced in 1887 from Australia to combact the cottony cushion scale *Icerya purchasi* in the nascent California citrus industry. The huge success of this project became an example of the great potential of classical biological control as a tactic for suppressing invasive pests (DeBach, 1964).

Subsequently, the beetle was introduced to at least 29 other countries including India and control of the target was either complete or substantial in all locations. The success of the vedalia beetle catalyzed the introduction of many exotic coccinellids targeting other pests, although rarely with the same degree of success (Caltagirone and Dout, 1989). The high degree of specificity of *R. cardinalis* for *Icerya purchasi* species in combination with a very short generation time, highly efficient detection of isolated host patches, and the ability of a larva to complete development on a single mature female scale are all though to be key factors in the effectiveness of the vedalia beetle (Prasad, 1990).

Diomus hennesseyi and *Hyperaspis notate* against *Phenococcus manihoti*

Against the cassava mealy bug, *Phenococcus manihoti,* the neotropical ladybirds, *Diomus hennesseyi, Hyperaspis notate* were both established in central and eastern Africa but none of those gave significant success against the target pest (Neuenschwander, 2003).

Hippodamia convergens against *Schizaphis graminum*

Introduction of a US strain of *Hippodamia convergens* to Kenya for control of *Schizaphis graminum*in 1911 failed despite a good history of the predator-prey association on wheat (Hunter, 1909). This was probably because the source material was not adapted to tropical conditions (Greathead, 1971).

Cryptolaemus motrouzieri against Mealy Bug

The most notable ladybird predator of mealy bug is *Cryptolaemus montrouzieri* Mulsant. Great head (2003) reviewed classical programmes in Africa and listed nine coccinellid species successfully established, of which eight were scale feeding species and the other mealy bug predator *Cryptolaemus montrouzieri* (Figure 11).

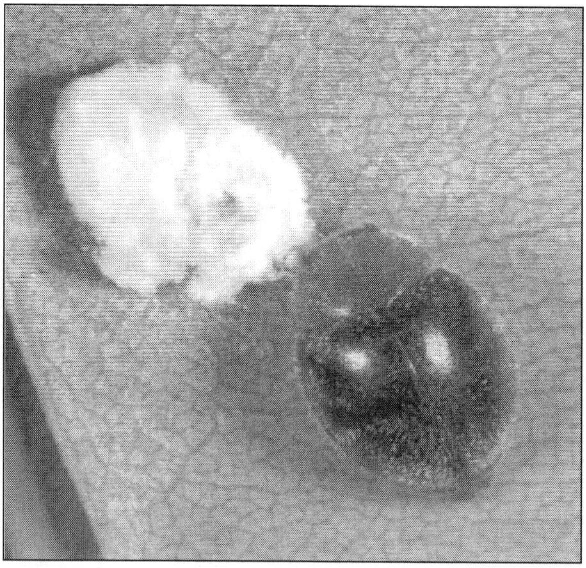

Figure 11: ***Cryptolaemus motrouzieri*** **Feed on Mealy Bug.**

Cryptognathanodiceps sp. against *Aspidiotus destructor*

Successful control of the coconut scale, *Aspidiotus destructor,* was attributed to the introduction and establishment of *Cryptognathanodiceps* sp. Marshall from Trinidad to Fiji (Singh 1976).

Rhyzobius lophanthae against Diaspidinae Scale

Successes have been achieved by introducing *Rhyzobius lophanthae* in controlling various scales under the subfamily Diaspidinae in Italy, the Black Sea coast of the Ukraine, and North Africa (Yakhontov 1960). The species is now considered an important introduced predator of citrus scale insects in both the United States (Flint and Dreisdadt 1998) and Australia (Smith *et al.,* 1997)

Chilocorus circumdatus against Citrus Snow Scale, *Unaspis citri*

Chilocorus circumdatus, has been successfully introduced for controlling citrus snow scale, *Unaspis citri* to Australia, South Africa, the USA and elsewhere (Samways *et al.,* 1999). It appears quite specific to its prey and is able to track it effectively at low densities, although in Hawaii it reportedly feeds on two additional species of Diaspididae (Funasaki *et al.,* 1988). *Chilocorus circumdatus* is now considered to be the most important predator of *U. citri* in Australia (Smith *et al.,* 1997).

Chilocorus bipustulatus against Armoured Scale Insects

Successful introduction of *Chilicorus bipustulatus,* an effective predator of armoured scale insects was achieved in USA (Huffaker and Doutt. 1965), North Africa (Iperti and Laudeho, 1969), central Africa (Stansly, 1984) and Australia (Waterhouse and Sands 2001).

Chilocorus nigritus against Variety of Armoured Scales

Chilocorus nigritus feeds on a variety of armoured scales and has been widely disseminated throughout tropical regions through both intentional and unintentional introductions (Samways, 1989). As early as 1940 it was introduced effectively against a complex of scale insects damaging coconuts in the Seychelles (Greathead, 2003).

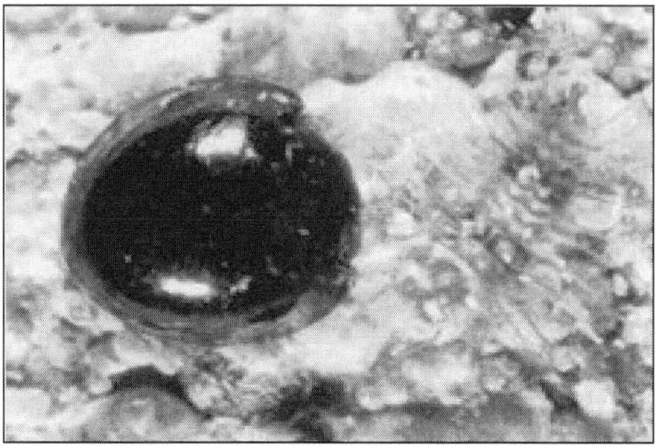

Figure 12: *Chilochorus nigritus* Feeds on Coconut Armoured Scale.

Aphidecta obliterate against Balsam Woolly Adelgid

The larch ladybird, *Aphidecta obliterate,* a species of European origin was introduced to South Carolina in 1960 (Amman 1966) to control the balsam woolly adelgid, *Adelgespiceae,* an invasive pest that arrived in North America around 1900. Later, *A. obliterate* was introduced to British Columbia, Canada (Harris and Dawson 1979) where it has emerged as an important biological control of both *A. piceae* and the hemlock woolly adelgid, *Adelges tsugae* (Humble 1994, Montgomery and Luon 1996).

Cheilomenes sexmaculata, Coccinella transversalis, C. septumpuntata, against Aphids

Dixon (2000) mentioned, most of the predaceous coccinellids species are worldwide that specifically targeted aphids and judged only one to be 'substantially successful'. The high reproductive rates of aphids are achieved through a combination of *thelytoky* and a telescoping of generations. Consequently, multiple aphid generations can be completed within the time required for a single generation of any coccinellid. *Cheilomenes sexmaculata, Coccinella transversalis, C. Septumpuntata* are polyphagous. These all are predate more numbers of aphid species.

Sasjisscymnus tsugae against *Adelgis tsugae*

Sasjisscymnus tsugae, an Asian was introduced from Japan to target *A. tsugae* and has been established in various regions of the eastern USA due to release programmes initiated during 1997 (Cheah *et al.,* 2004). Although established, the impact of *S. tsugae* on the pest population was not as great it was expected.

Mite Feeders (Acarophages)

Among Coccinellidae, mite feeders belong exclusively to a single group classified in two genera, *Stethorus* and *Parastethorus.* Fleschner (1950) showed that the larva of *Stethorus* has grooved mandibles that are used for extraoral digestion.

Mychophagy

All species of this group feed on conidia and hyphae of powdery mildews, fungi classified as Erisyphales (Ascomucota). This group is well represented in South America, Europe and Asia but in Australia limited to two species of genus *Illeis.*

Feeding on Plants

The sub family Epilachininae is cause considerable damage to various plants specially belonging to the families of Cucurbitae, Solanaceae and Papillionaceae, often they completely defoliate the plants like Bitter gourd (*Momordica charantia* L.), Karkole (*Momordica cochinenchinensis* Speng) and cause serious damage to Brinjal (*Solanum melongina* L.), Tomato (*Lycopersicum esculentum* Mill) and Potato

(*Solanum tuberosum*). The plant feeding species are *Epilachna vigintioctapunctata* Fab., *E. septima* Dieke, *E. dodecastigma.*

Figure 13: Ladybird Beetle Feeding Pollen.

Alternative Food

Lady bird larvae and adults may supplement their normal prey in times of scarcity with other types of food. They consume flower nectar, water and honey dew (the sugary excretion of piercing and sucking insects such as aphids and whiteflies. Many plants species also contain organelles in locations on the plant other than the flower termed as **extrafloral nectaries**, that produce nutrient. While it was first thought that extrafloral nectaries were used by the plant for excretion, it is well substained and that most plants actually use the extrafloral nectaries to attract predators and parasites for protection from their herbivores. Extrafloral nectaries may be located on leaf laminae, petioles, rachids, bracts, stipules, pedicels, fruit *etc.* Ladybird beetle often use the secretion from extrafloral nectaries in their diet and are just of the many beneficial insects that use extrafloral nectar secretions.

Chapter 9

Ladybird Beetles in Predatory Guilds

Heterospecific predators, including ladybirds, combat for a common food resource habitat (guild). As the common food (extraguild prey) diminishes they start attacking each other. In this struggle, one becomes dominant (intraguild predator) and overpowers the other (intraguild prey). Intraguild interactions are not the simplest and may give rise to series of complex interactions, which could both be advantageous and disadvantageous to the fitness of ladybirds and to the fate of biocontrol.

Isolation of the Guild Members

Occasionally, intraguild members share common food and space resources but they are isolated and rarely bother each other. Few ladybirds differ spatially in the guild by sharing different lateral or vertical spaces in prey habitat (Flowers *et al.*, 2007). Plant morphology and other characteristics may also act as a barrier between the intraguild members (Janssen *et al.*, 2007). This spatial isolation of intraguild predators might also have a mutual effect on biocontrol, if a predator facilitates its other spatially isolated guild member in fetching itself to the food resource (Losey and Denno, 1998). Ladybirds also have a difference in the temporal distribution and are isolated in different time zones (Toda and Sakuratani, 2006).

Ladybirds in the Guild

Ladybirds involved in intraguild combat are more conspicuous if the extraguild prey is a limiting factor. They exploit the food resource by attacking the extraguild prey and reducing their abundance for their competitors. This exploitative competition of food resource is more advantageous to the smaller-sized ladybird as they have lesser food demands (Obrycki *et al.,* 1998). As the density of extraguild prey decreases, frequency of intraguild predation (IGP) increases (Gagnon, 2010). On the contrary, an increase in extraguild prey density creates a dilution effect which manages the IGP and subsequently lowers its possibilities (Lucas and Brodeur, 2001). IGP is responsible for habitat dislocation (including both emigration and immigration) of intraguild members and affects the biocontrol events. An increase in the density of intraguild prey also increases the frequency of IGP, as the probability of encounters with the competitors will also increase (Noia *et al.,* 2008). IGP is directed towards the main intraguild predator as the status of intraguild predator and prey will never change regardless of differences in the extraguild prey and its habitats (Yasuda and Ohnuma, 1999).

The relative size and stage, mobility of species, aggressive strategies, mandibular structure, degree of feeding and habitat specificity, defense strategies, and density of extraguild prey determine the outcome of IGP (Yasuda *et al.,* 2001; Lucas, 2005; Roy *et al.,* 2006). Disparity in body size also increases the likelihood of IGP and it skews towards larger and voracious species. Even their lower instars attack higher instars of inferior and less voracious species. Frequency of encounters of same-stage instars is usually lower. However, the older instars frequently encounter each other more than the younger ones (Yasuda *et al.,* 2004). IGP also provides nutritional benefits to the developing stages if the extraguild preys lack nutritional values (Ware *et al.,* 2009).

Interaction within Ladybirds

The majority of ladybirds attack, preys upon, and displaces the other members of their family in a limited food resource struggle. Invasion and establishment of aggressive species following displacement of indigenous ones may be an outcome of IGP. One such invader is the Harlequin ladybird, *H. axyridis.* This invader has a competitive advantage over indigenous species, such as *C. septempunctata* (Yasuda *et al.,* 2001), *Coleomegilla maculata* DeGeer (Labrie *et al.,* 2006), *Hippodamia variegata* Goeze, and *A. bipunctata* (Lanzoni *et al.,* 2004) due to its vast prey range, higher predation, and foraging potential (Alhmedi *et al.,* 2010a). It is acclaimed to be the most invasive ladybird on Earth (Roy *et al.,* 2006). This, along with other successful invaders, may affect the dynamics and composition of indigenous species guilds. This invasive species indulges in interference competition resulting from direct interaction of competitors or from the modification of a resource by one competitor such that it becomes unusable by another. Harmonia axyridis is acclaimed as top predator because of magnitude, direction, and symmetry of IGP (Felix and Soares, 2004).

There is, however, no reported IGP between adults of two species; still the adults of *H. axyridis* may interact negatively by interfering with each other's foraging and oviposition capabilities. They even interfere with other competitors when food resource is not undergoing shortage (Soares *et al.,* 2008). The dominant species, especially *H. axyridis*, also indulge in asymmetric reproductive interference that may compel the inferior indigenous species to become specialist predators of less preferred prey in nature (Noriyuki *et al.*, 2012).

The invasive species also gets benefitted by apparent competition, results from the indirect interaction of two co-occurring species sharing a common resource. Adult *H. axyridis* have the intrinsic advantage of being the least vulnerable to parasitism by braconid parasitoid, *D. coccinellae*, and fungus, *Beauveria bassiana* (Balsamo) Vuillemin (Roy *et al.*, 2008) compared with other co-occurring species. The vulnerability of other ladybird species toward parasitism negatively affects their fitness, decreasing their competitive ability. Apart from being a victor in competitions, *H. axyridis* is the biggest winner in almost all IGP interactions within the family. Its eggs are more chemically defended thereby skewing IGP of eggs in its favor (Cottrell, 2007). Morphological traits, like larger body size, the presence of spines on the back of the third and fourth instars provides extra protection (Ware *et al.*, 2008). Additionally, size, strength of the integument, and distastefulness of its pupae reduce its vulnerability (Felix and Soares, 2004). Larvae of this invasive species are larger and more aggressive than those of other ladybirds (Yasuda *et al.*, 2004). When paired with a larger larva of a different species, *H. axyridis* can either win (Michaud, 2002) or lose (Ware and Majerus, 2008). Other ladybirds that fall victim to *H. axyridis* may be superior and top intraguild predators in their respective guilds (Rondoni *et al.*, 2014).

Interaction with Other Predators

Ladybirds also struggle with other group of insects outside the family for food resource. They usually co-occur with chrysopid (Neuroptera: Chrysopidae) larvae and share limited resources. However, IGP interactions between them are asymmetrical which could have positive, negative, or neutral impacts on aphid biocontrol (Gardiner and Landis, 2007). *Harmonia axyridis* mostly acts as an intraguild predator and declines local populations of chrysopids and predatory gall midge along with that of aphids (Gardiner and Landis, 2007). A few top ladybird species fall victim to voracious chrysopid larvae, especially to *Chrysoperla carnea* (Stephens) (Noppe *et al.*, 2012) despite being equipped with strong defense. Ladybird larvae wriggle and bite and may use armor of their dorsal and lateral spines (Michaud, 2003) to attack the enemy. Some ladybird pupae flip when perturbed which may sniff off the antennae of the attackers (Lucas *et al.*, 2000). Both chrysopids and ladybirds use chemical defenses. Coccinellid larvae (Lucas, 2005), pupae (Lucas *et al.*, 2000), and adults all use chemical substances containing alkaloids to deter predators. Ladybirds could be victimized by other predators, *e.g.*, big instars of syrphids, pentatomids, and spiders in the guild.

Chapter 10

Important Coccinellid Predators in different Ecosystems

Rice

The Coccinellid fauna in the terai region of West Bengal were identified as *Coccinella septumpunctata* Lin. and *Micraspis sp.* (Ghosh *et* at., 2013). Numerous reports show Coccinellids as opportunistic, feeding on a variety of food material in addition to their preferred prey. *Micraspis discolor* is the most abundant species of Coccinellid in rice ecosystems and touted as a biocontrol option for brown plant hopper (BPH), *Nilaparvata lugens* (Stal), a key pest of rice. However, it has been reported as both entomophagous and phytophagous. The abundance of adult *M. discolor* in rice at flowering phase does not correspond to prey abundance in the field but rather reflects an inclination to pollen feeding more than entomophagy (Shanker *et al.*, 2013).

Ten genera under four tribes of the family Coccinellidae were collected and identified from Indo Bangladesh border in north-east India (Chowdhury *et al.*, 2015). They are *Brumoides suturalis* Fabricus), *Chilocorus nigritus* (Fabricus), *Coccinella septumpunctata* (Lin.), *C. transversalis* (Fab.), *Harmonia octamaculata* (Fab.), *Illeis cincta* (Fab.), *Micarspis crocea* (Fab.), *Propylea sp. Nr. Japonica* (Thunberg), *Cryptolaemus montrouzieri* Mulsant and *Scymnus* (*Scymnus*) *nubilus* Mulsant.

Result also revealed that out of ten coccinellid *Harmonia octomaculata* (Fabricius) (22.9 per cent), *Micraspis crocea* (Fabricius) (18.5 per cent), *Brumoides suturalis* (Fabricius) (12.36 per cent) and *Chilocorus nigritus* (Fabricius) (9.06 per cent) were major predatory coccinellids beetles in rice ecosystems of Indo-Bangladesh border in the north-eastern region of India.

Wheat

Coccinellid fauna found in whaet crop was *Coccinella transversalis* Fabricus and *Cheilomenes sexmaculata* (Fabricus) (Megha *et al.,* 2015, Jat *et al.,* 2009) also noticed the population of *C. sexmaculata* in wheat crop.

Maize

Coccinellid fauna found on maize crop was *Scymnus nobilis, Harmonia ocatmaculata* and *Cheilomenes sexmaculata* (Meha *et al.,* 2015). The predatory activity of *Cheilomenes sexmaculata* in maize was noticed by Tank *et al.,* 1999. The activity of *Scymnus nubilis* in maize has not been reported by earlier authors.

Sorghum

Coccinellid species found in sorghum crop were identified as *Cheilomenes sexmaculata, Illeis cincta* and *Harmonia octamaculata* (Patil and Sathe, 2003). However, *H. octamaculata* has been recorded for the first time which has not been reported earlier.

Mustard

The Coccinellid species namely *Coccinella septumpunctata* Lin., *C. transversalis* Fab., *Menochilus sexmaculatus* Fab., *Micraspis discolor* Fab., *Coelophora unicolor* Fab. And *Oenopia luteopustulata* Mulsant have been recorded in the terai region of West Bengal predating on the mustard aphid (Choudhury and Pal, 2006)

Cowpea

Coccinellid species found in cowpea crop were identified as *C. trasversalis, Cheilomenes sexmaculata, H. octamaculata, H. variegate, Brumoides suturalis* and *Rodalia cardinalis* by (Debraj ann singh, 1990), (Agrwal and Bhattacharya, 1995), Joshi *et al.,* 1999), Chanmala (2009), *H. octamaculata* by Chanmamla (2009) on cowpea crop were reported. However, *B. suturalis* and *R. fumida* on cowpea crop have not been reported earlier which forms the new report.

Green Gram

Coccinellid fauna found in green gram was *C. sexmaculatus, C. transversalis* and *I. cincta.* The present findings are in line with the findings of (Sharma *et al.,* 1993) who reported that the adults of *C. transversalis* were reorded along with *A. craccivora* infesting green gram. *Illeis cincta* and *C. sexmaculatus* has not been recorded on green gram by previous workers.

Safflower

Coccinellid species found in safflower crop were identified as *H. variegate* and

C. sexmaculatus (Joshi *et al.,* 1999), (Kamath and Hugar, 2001) alos reported the predatory acticity of *C. sexmaculata* and *C. septumpunctata* on safflower.

Groundnut

Coccinellid fauna found in ground nut were *C. transversalis, C. sexmaculata* and *H. variegate*. Bakhetia and Sidhu (1997), who reported that *M. sexmaculata* was found to to be the most active during August and September in ground nut, Mathew *et al.* (1971) and Patel *et al.* (1976) and Chanmala (2009) reported the activity of *C. transversalis, C. sexmaculata* and *H. variegata* in ground nut.

Soybean

Coccinellid fauna found in soybean were *C. sexmaculatus* and *C. transversalis.* The present findings are in accordance with the findings of Tank (2006) who reported that no grubs of the coccinellid predator appeared on soybean crop, however negligible and scattered numbers of beetles were seen.

Cotton

Coccinellid species found in cotton crop were identified as *C. sexmaculatus* and *I. cincta*. The present findings are in accordance with the findings of Tank *et al.* (2008), Varma *et al.* (1993), Veeravel and Baskaran (1995), Joshi *et al.* (1999), Singh and Sharma (2010) also reported the activity of *C. sexmaculatus* on cotton crop.

Bhendi

The coccinellid species found in Bhendi crop was identified as *I. cincta.* Similar results were obtained by Tank *et al.* (2008), who reported the activity of *C. sexmaculata* on okra.

Lucerne

The coccinellid species found in Lucerne crop were *C. transversalis, C. sexmaculatus, H. octomaculata, H. variegata, and Propylea dissecta.* Tank (2006) also noticed the activity of grubs and adults of *C. sexmaculatus* on Lucerne crop. However, the activity *C. transversalis, H. octomaculata, H. variegata, and P. dissecta* in Lucerne have not been reported earlier.

Mulberry

The coccinellid species found in mulberry crop were identified as *I. cincta* and *Illeis* sp. These coccinellids were observed to feed on powdery mildew, *Phyllactinia corylea* (Pers). Krishnakumar and Maheshwari (2004) also reported the activity of above two species.

Cabbage

The coccinellid species found in cabbage crop were identified *C. sexmaculatus and H. variegata*. The predatory activity of *H. variegata* was reported by Sharma and Verma (1993) who also reported that adults of *Adalia variegata* were the first to resume activity on cabbage in Himachal Pradesh. Mandal and Patnaik (2008) reported the activity of *C. sexmaculata* during November to February.

Table 1: Some Coccinellid Beetles with their Hosts

Name of the Species	*Host Plants*
Jauravia pallidula	Citrus and Som
Cryptogonus bimaculatus	Citrus and Som
C. quadriguttatus	Citrus
Brumoides suturalis	Chilli, Brinjal, Okra, Cowpea and Wheat
Anisolemnia dilatata	Som
Coccinella septempunctata	Wheat, Mustard, Brinjal, Chilli, Cabbage, Cowpea, Citrus and Som
C. transversalis	Brinjal, Chilli, Okra, Cabbage Cowpea, Rice, Wheat, Maize, Mustard and Linseed
Cheilomenes sexmaculata	Rice, Wheat, Maize, Linseed, Chilli, Okra, Cowpea and Som
Harmonia octomaculata	Wheat
H. dimidiate	Wheat, Brinjal and Som
Micraspis discolor	Rice, wheat, Bhendi, Brinjal, Chilli and Mustard
M. yasumatsui	Rice, wheat, Bhendi, Brinjal, Chilli and Mustard
Propylea dissecta	Wheat, Brinjal, Chilli and Cowpea
P. japonica	Rice, wheat, Brinjal and Som
Illeis indica	Brinjal, Chilli, Cowpea and Mulberry

Chilli

In chilli crop, the coccinellid species found were identified as *I. cincta* and *C. sexmaculatus*. (Tatagar *et al.*, 2012) reported the activity of *C. sexmaculata* in chilli crop. The activity of *I. cincta* has not been reported by earlier workers.

Bitter Gourd

The phytophagous coccinellid species observed in bitter gourd was *Epilachna vigintioctopunctata* (Fabricius). Deshmukh *et al.* (2012) reported the activity of *E. vigintioctopunctata* in bitter gourd which agrees with the present findings.

Brinjal

The phytophagous Coccinellids species observed in brinjal was *E. vigintioctopunctata*. The present findings were in agreement with the findings of Savita and Anandhi (2006), who reported the activity of *E. vigintioctopunctata* in brinjal crop.

Tea

Twenty species of Coccinellid predators were observed in Sub-Himalayan tea plantation of North Bengal (Roy *et al.*, 2010). Of these, *Micraspis discolor* (Fab.)

was dominant (42.5 per cent) in the conventionally managed tea plantations. The abundance of *Micraspis discolor* populations was positively correlated with the abundance of red spider mites (*Oligonychus coffeae* Neitner, Acarina: Tetranychidae) and tea aphid (*Toxoptera aurantii* Boyaer de Fons, Homoptera: Aphididae). Das *et al.* (2010) also observed the Coccinellid as the dominant group of predator in the tea planations of North Bengal with 27 reported species.

The species diversity in the Horticulture ecosystem was more as compared to Agricultural ecosystem and Organic farming ecosystem.

Chapter 11

Ladybird Beetle as a Biocontrol Agent

Biological control is the use of animals, fungi, or other microbes to feed upon, parasitize or otherwise interfere with a targeted pest species. Successful biocontrol programs usually significantly reduce the abundance of the pest, but in some cases, they simply prevent the damage caused by the pest (*e.g.* by preventing it from feeding on valued crops) without reducing pest abundance (Lockwood 2000).

Biocontrol is often viewed as a progressive and ecofriendly way to control pest organisms because it leaves behind no chemical residues that might have harmful impacts on humans or other organisms, and when successful, it can provide essentially permanent, widespread control with a very favorable cost benefit ratio. However, some biocontrl programs have resulted in significant, irreversible harm to untargeted (non-pest) organisms and to ecological processes. Of course, all pest control methods have the potential to harm non-target native species, and the pests themselves can cause harm to non-target species if they are left uncontrolled. Therefore, before releasing a biocontrol agent (or using other methods) it is important its potential to benefit conservation targets and management goals against its potential to cause harm.

Types of Biological Control

Organisms used to feed on, parsitize, or otherwise interfere with targeted pests are called biocontrol agents. There are several general approaches to using biocontrol agents:

1. Manipulative biological control.

2. Augmentative biological control
3. Classical biocontrol

Use of Ladybirds in Biological Control

Most species of ladybirds are considered beneficial because they are predators of Homoptera or Acarina, many of which are considered to be pests. These predatory ladybirds contribute to the regulation of populations of their prey, and in some situations contribute a high level of regulation. When ladybirds naturally contribute a high level of control of pests, or in combination with other predators and/or parasitoids and diseases contribute a high level of population regulation of pests, people may benefit. That is to say that farmers may benefit, at no cost, because they have no or negligible pest problems.

Sometimes, farmers mistake the ladybird larvae for pest and spray chemical pesticides that kill them. The result is increased problems from real pests. The answer is a constant educational effort to inform people about ladybirds and what their larvae look like. This effort cannot be end, because people knowing nothing about ladybird life cycles are born each minute.

Manipulative Biological Control

The objectives are:

- ☆ To capitalize on the ladybirds that already are present,
- ☆ To make conditions as favorable as possible for them (Manipulation)
- ☆ And especially to avoid spraying chemicals (Insecticides, fungicides, or herbicides) that will harm them. (Conservation)

Augmentative Biological Control

The term 'augmentation' refers to the periodic release or inoculation of a natural enemy in contexts where only the released insects or their immediate descendants are expected to exert biological control. Recent reviews of the augmentation approach continue to indicate that coccinellids are rarely utilized in such programmes. Two limiting factors are the cost of production and the education required their effective application by farmers. Thus many demonstration projects have shown potential efficacy, but have failed to evolve into tactics that are attractive or economically viable pest control alternatives.

This begins with the recognition that ladybirds in a given pest situation are present but too few to do the job required, and buying more from a commercial producer like ICAR institutes, SAUs to release to augment those already present.

Life Stages for Release

A risk is that if adult ladybirds are released, many of them fly away. But, if ladybird larvae released, they have the option of eating the pest with which they are presented, or starving. They cannot fly away. Obviously, this requires matching the pest to a purchased ladybird species that will eat that pest otherwise it will be waste.

Selection of Source Material for Augmentation

Care should be taken in extrapolating control potential from laboratory feeding experiments, because the range of prey acceptable to coccinellids in confinement is often much broader than that in the field. This can be true for various reasons, including prey specificity, plant and habitat preferences or simple refuge effects that are eliminated in simplified feeding situations. The export of mass collected beetles for augmentation in dissimilar habitats (where they are often released against unfamiliar prey) may be misguided, even when locally indigenous species are involved. Such export is based on the implicit assumption that populations of a species are ecologically uniform across their range and ignores the likely importance of local adaptations in providing effective biological control.

Classical or Inoculative Biological Control

This type of biological control, some individuals of a ladybird species that is not already present are released in the hope that they will establish a population and eventually control the pest that is of concern. Classical biological control typically applies to a situation in which a new pest has invaded, and researchers import and release a ladybird that is believed to control the pest elsewhere. Typically, the imported ladybird becomes established or does not become established; if it becomes established, it may or may not control the pest in thus new situation.

Greenhouses provide a habitat for plants and pests and biological control agents that differs from outdoor habitats. Typically, culture begins with initiation of a crop of plants that has no pests (or seems to have none). But then, pests somehow show up, and there are no ladybirds (or other organisms) to control them. The situation is very much like that of classical biological control, and ladybirds released into the greenhousesmay control the pests and eliminate need to the chemical pesticides.

Chapter 12

Conservation of Ladybird Beetles

For most agricultural system, conservation techniques for coccinellids are lacking, even though they are abundant in these habitats. Conservation of coccinellids can be facilitated by modifying the environment so that minimal disruption can be caused to the coccinellids. In recent years, environmental concerns have given a renewed emphasis to the preservation and enhancement of indigenous coccinellid species in order to improve their bio control contributions in natural and managed ecosystems. These are a wide variety of tactics for conserving coccinellids in horticultural and agricultural settings.

Conservation of natural enemies is an important component of pest management, which can improve their efficacy against target pests. Conserving predaceous Coccinelidae species in agricultural ecosystems is used to enhance their biocontrol contribution. Favourable conditions in these habitats can contribute to a more efficient population regulation of several pests. Conservation efforts focus on discouraging emigration from a crop system and enhance retention time of coccinelids in periods with low prey availability. Thus, the management of agroecosystems should focus on providing resources in such temporal and spatial scale that may prevent their emigration or attract them in habitats. In addition, in a greenhouse ecosystem, another conservation action is to reduce mortality and sublethal effects caused by insecticides.

In general, these aim to encourage immigration, discourage emigration, mitigate mortality caused cultural activities including the application of pesticides, or improve survival during critical periods by the provision of specific resources, usually in the form of particular plant species.

Plants may directly or indirectly influence coccinellid effectiveness as a result of altered prey suitability or host finding success of the predator. Prey reared on resistant host plants may have a significant impact on the fitness of coccinellids. However, because these effects are generally complex and interacting, generalities about the compatibility of plant resistance and biological control are lacking (Obrycki and Kring, 1998).

Changes in cultural practices can be achieved through maintaining refugia for coccinellid beetles by adopting strip planting, field borders, eco-feast cropping and cover cropping. Agricultural practices that use conservation tillage and cover crops are promoted for their benefits to soil quality, weed suppression and water conservation (Schomberg, 2003). Also they are noted for their ability to encourage biological control, with numerous cases where cover crops have been shown to increase beneficial species and reduce pest species (Olson *et al.,* 2012).

The abundance and efficacy of coccinellid as biocontrol agents may depend on the structure, composition and diversity of the surrounding landscape at different spatial scales. The worldwide trend toward simplification of agricultural landscapes, caused by planting annual crops and eliminating edges of less disturbed natural or semi-natural habitats, can alter the structure of natural enemy communities, favoring exotic over native species and lowering the diversity of natural enemies. This, in turn, could result in a lower biological control because of the loss of species with different roles in the agro-ecosystems (Grez *et al.,* 2014).

The non-farmed areas surrounding agricultural lands are therefore an important component of the stability of agro-ecosystems. These areas can be putative refuges for predators and can act as an important source of biological control agents. Refuges can enhance the survival, fecundity, longevity, and behavior of natural enemies, increasing the colonization of crops by such enemies and contributing to the control of the pest populations. Ecologically based research on biological control by natural-enemy populations require an understanding of the population dynamics of these species and of their dispersal into the crop from a particular resource or habitat patch.

Coccinellids are an important group of predators that should be conserved in the different crop ecosystems. The rate of predation and species diversity of coccinellid community is influenced by plant community structure and micro climate (Tooker and Hanks, 2000). This is because of different species of ladybird exhibit a preference for specific vegetation types coupled with suitable food in sufficient abundance.

Rearing and Release Methods

One focus for rearing imported coccinellids has been on the effects of temperature and prey on development and reproduction. Symbiotic microorganisms may cause female-biased sex ratios in coccinellids. Effects of inbreeding during rearing also need to be considered relative to establishment. Importation programs using coccinellids have historically released as many adults as were available from quarantine and rearing. Introduction of large numbers of predaceous adults in a single location may cause density-dependent movement. Furthermore, many coccinellids species are cannibalistic, and competition for prey may occur.

Habitat Management

Conservation of coccinellid populations hinges on understanding the full range of resources utilized by all life stages of a coccinellid species throughout a complete annual cycle. Potentially important considerations include the preservation of natural shelter or overwintering sites, the diversification of plant communities to improve availability of food and shelter, and modifications to conventional agricultural practices that minimize impacts on coccinellid populations. These modifications include tactics such as strip harvesting, intercropping, reduced tillage, restriction of pesticide application to spot or strip treatments, and the use of pesticides with selective modes of action.

Strip-Harvesting

Strip harvesting alfalfa (lucerne) is one way to prevent the post- harvest emigration of local predator populations as a consequence of food deprivation, thus conserving natural enemies in fields and facilitating their rapid colonization of harvested areas when re growth occurs. This approach has long history of use in alfalfa production (Scholl and Medler 1947) where it functions to stabilize biological control of a range of pests and conserve coccinellids in other perennial forage crops where aphids may cause losses.

Figure 14: Strip Cropping.

Floral Diversity

Coccinellids had consistent patterns of occurrence on particular non-crop plant species, mostly common weeds, and avoided others. These affinities were often independent of the presence of the prey as fully 40 per cent of the coccinellids were observed on plants without aphids. Border vegetation planted provides supplementary food resources or alternative prey for coccnellids may sometimes impede their timely dispersal into adjacent crops, a phenomenon often referred to as 'apparent competition'.

Intercropping

Sorghum with cotton could increase some species of ladybird populations in cotton due to the abundance of green bug in sorghum early in the growing season.

Reduced Tillage

Reduced tillage or no tillage decreased aphid colonization in wheat and sorghum, due to crop residues functioning as mulch with reflective properties that reduce the apparency of the crop plant. There also potential benefits for natural enemies such as coccinellids, the predators most often identified as responsible for biological control of cereal aphids.

Alternative or Supplementary Food

The rationale for providing additional food for coccinellids is that food can be limiting to coccinellids survival or retention in the crop during critical periods, or can be used to attract immigrant beetles into crop fields earlier in the season or in larger numbers. The assumption is that provision of food can alter the spatial distribution of coccinellids in the landscape so as to improve subsequent biological control of a pest that is initially present at very low levels, or one that arrives in a crop with some predictability with respect to season or plant growth stage.

Artificial honeydews have been employed successfully to attract a variety of aphid predators, including coccinellid, into cropping systems. Many plant species also contain organelles in locations on the plant other than the flower termed as extarfloralnectaries which is produce a nutrient laden secretion. While it was first thought that extrafloralnectaries were used by the plant for excretion, it is well sustained that most plants actually use the extrafloralnectaries to attract predators and parasites for protection from their herbivores. Over 2000 species of plants in 64 families have extrafloralnectaries.

Use of Herbivore-Induced Plant Volatiles

Plants response to attacks by herbivores by releasing volatiles. These herbivore-induced plant volatiles (HIPV) have the potential to attract beneficial arthropods. Several HIPV compounds and synthetic plant volatiles has been tested in laboratory and field conditions and supports this evidence. For instance, methyl salicylate is known to attract predaceous Coccinellidae species in several agricultural habitats, acting as an indirect defence mechanism of cultivated plants. Prey monitoring and time of application are key components for the successful use of synthetic plant volatiles.

Hibernation Refuges

In temperate regions where overwintering mortality can be high, the planting of hibernation refuges such as 'beetle banks' has successfully enhanced coccinellid survival in various agricultural contexts. Artificial hibernation refuges for coccinellids in arboral settings. Construction of artificial shelters may sometimes be feasible in high value vegetable crops, but is probably cost- prohibitive in most large scale agricultural settings and current approaches to enhancing overwintering survival tend to rely on recognizing and preserving natural hibernation sites.

Selective Use of Pesticides

This has long been recognized as an important tactic for conserving coccinellids, and efforts have been made in recent years to determine indices of relative toxicity that can be used to rank materials for compatibility with ladybirds in IPM programmes. In general, eggs and young larvae are the most sensitive life stages and adults the least. As sessile stages, eggs and pupae may have a somewhat reduced risk of exposure in the field relative to actively foraging larvae and adults.

One problem in assessing the impact of insecticides in field studies is that materials highly effective against the pest population will also reduce numbers of coccinellids as a consequence of eliminating their food supply, even if they lack direct toxicity to the beetles (Poehling and Dehne 1984, Mateeva *et al.,* 2001).

However, careful timing of insecticides applications relative to crop development can sometimes serve to synergize, rather than disrupt, pest control by a complex of natural enemies that includes coccinellids (Flagan *et al.,* 2010).

Coccinellids are more tolerant of pesticides than their prey. More recently, Gerasha (2007) found that Pirimicarb, Imidacloprid and Thiamethoxam were all less toxic to *C. undecimpunctata* than to its prey, *B. brassicae,* although only adult beetles were tested. Broad-spectrum pyrethroids tend to be more destructive to coccinellids than organophosphates (Kumar and Bhatt 2002), although there is evidence that populations of *Hippodomia convergens* can evolve substantial resistance to pyrethroids in agro ecosystems with heavy usage patterns (Ruberson *et al.,* 2007).

Spinosad is generally benign to coccinellids, although Galvan *et al.* (2005) found sub-lethal impacts on development and reproduction. Because spinosad requires ingestion to induce toxicity, residue trials typically yield high safety ratings, whereas direct topical applications may result in some mortality, probably due to ingestion via grooming behavior (Michaud 2002). Although potentially hazardous to Hymenoptera, certain spinosad-based fly baits have tested relatively safe for cocinellids (Michaud 2003), possibly because the formulation did not stimulate consumption.

Pymetrozine selectively inhibits aphid feeding behavior, has reputedly low toxicity to coccinellids, and has been used effectively in conjunction with *C. septumpuntata* for control of *B. brassicae* on broccoli (Achermpong and Stark

2004). However, despite observing low acute toxicity of Pymetrozine to *C. leonine* and *C. transversalis*, Cole *et al.* (2010) found that less than 3 per cent of exposed larvae survived to maturity. Insecticides toxic to coccinellids may also be selective by virtue of formulation or mode of application. Systemic materials applied as granules, soli drenches or planting time 'in furrow' applications can give good control of aphids without causing direct mortality to coccinellids.

Most recently, the modern Neonicotinoids Thiomethoxam and Imidacloprid display good systemic activity in plants and have become increasingly available as seed treatments that are very effective against seedling pests, especially aphids and flea beetles, while sparing coccinellids and other predators from direct impacts, except when they engage in herbivory Moser and Obrycki 2009). Another systemic Neonicotinoids, Dinotefuran, was shown to provide good control of armored scales on Christmas trees when applied in a band around the base of the trunk, without impacting the foraging activities of the scale predators *Chil. stigma* and *Cybocephalus nipponicus* (Cowles 2010). However, coccinellids can be impaired behaviorally or reproductively by non-lethal concentrations of insecticides that they may a require when feeding on contaminated pollen or nectar or contaminated prey (Singh *et al.*, 2004, Eisenback *et al.*, 2010).

Insect growth regulators (IGRs) such as Buprofezen and Pyriproxiyfen generally lack acute toxicity to the coccinellids, but may impair development (Hattingh and Tate 1995) and fecundity (Olszak *et al.*, 1994). Nevertheless, good knowledge of insect ecology can sometimes enable the judicious use of these compounds without disrupting coccinellids populations. Despite the high sensitivity of *Rodalia cardinalis* to IGRs, and residual activity longer than 6 months under California conditions, IGRs can be used to control pyrethroid resistant scale insects in citrus, provided that applications are delayed until after *R. cardinalis* has exerted control of *Icerya purchasi* in spring (Grafton-Cardwell *et al.*, 2006).

Future Directions

Predaceous Coccinellidae species contribute to the population regulation of several insect pests in greenhouse ecosystems. In future, this contribution could be advanced if conservation methods adopted in practice from growers. Thus, the cost of applying these methods -which in some cases is rather prohibitive- should be detailed. Also, alternative methods based on current knowledge on alternative food resources and hibernation refugees could be developed.

The use of HIPVs by predaceous coccinellids is considered as a promising topic for the conservation of this taxa in greenhouse ecosystems. The deployment of new synthetic HIPVs, in combination with new application methods, could lead to a more efficient biological control of plant pests. While several agricultural practices have been proposed for the conservation of predaceous coccinellids in several habitats, long-term quantitative analysis of coccinellids' population viability in such cases is lacking so far. Thus, a major question in this topic is not only to develop new conservation tools, but to optimize the current knowledge. In this task, population viability analysis based on quantitative conservation biology tools could lead to the evaluation and optimization of the proposed methods. This is considered as an important direction for future research.

Chapter 13

Mass Culturing of Ladybird Beetles

Most insect studies are dealing with laboratory-reared insects and most of these studies incorporate insects reared on artificial diets. In fact, the rearing conditions and the diet composition are critical parameters in any case. To get reliable coclusions from experiments, the nutrition of the insects tested has to be perfectly known and suitable, because the answers will depend on this probably the vision of Knipling E.F. (1979) is not completely realistic, but could be a goal for specific applications: "Mass rearing of insects is stiil a young science. Artificial rearing entails to fulfil nutritional needs but also the other physiological environment, respiration, excretion, especially for parasitoids (Grenier *et al.,* 1986).

The Mass Production for Practical Use

The mass production for utilization of predators or parasitoid insects in biological control strategies is the most evident aim. There are some advantages such as simiplification of the production line, higher flexibility of the production, reduction or suppression of allergy problems generated by scales, lower costs expected with scaling up effect but, insect colonies are also used by many commercial companies for pesticide testing, production of insect derivates. Of course there is a growing sector concerning the multiplication of beneficial insects. Espcially the bio-control strategy using inundative releases needs huge quantities of insects.

The basic food chain for entomophagous insect is the "natural" tritrophic system, with palnt, pest insect (host or prey), and predator or parasitoid. One of the first ideas to improve this production line was to try to reduce the number of compartments. A first step consisted in using bitrophic system, with two possibilities of sub-stitutes.

At last, the final solution to shorten the production line would be the availability of an artificial medium/diet allowing larval and pupal developments up to adult emergence, and possibly egg laying for predators.

Some of the mass production techniques for ladybird beetles have given below.

Mass Production of *Cryptolaemus montrouzieri*

C. montrouzieri has been introduced from Australia for the control of *Coccus viridis* on coffee. But the predator has established on many species of mealybugs and green shield scale. In the field its practical use for the suppression of mealybugs *viz.*, pink mealy bug, *Maconellicoccus hirsutus,* citrus mealy bug *Planococcus citri*, tailed mealy bug *Ferrisia virgata* and mealy scale *Pulvinaria maxima* on citrus, coffee, grapes and several other fruit crops and ornamentals has been demonstrated. Use of *C. montrouzieri* is the break through in applied classical biological control. The coccinellid predator is native of Australia. In 1892, it was introduced into California by Albert Koebele for the control of citrus mealy bugs. Following the success, the beetle was introduced into India in 1898 by New Port. It has given effective control of mealy bugs in fruit crops like citrus, grapes, guava, *etc. C. montrouzieri* is one of the outstanding examples in the biological control history. Its importance is also evident by its growing commercialization in India.

In the laboratory, the life cycle is completed in approximately 30 days. The premating and preoviposition periods are about 5 and 10 days respectively. The oviposition is about 10 days. Eggs are laid from late evening to early morning. They are pale yellowish white, the surface being smooth and shiny. It is oval to cylindrical, both the ends beings smoothly rounded. Incubation period ranges from 5 to 6 days but extended in winter months. Viability of eggs is 90 to 100 per cent. The newly hatched grub is sluggish but becomes active after 3 to 4 hours. The tiny grub is pale greyish with white lines across the body along intra segmental regions. These white lines become prominent after few hours and white wax strands develop after a day. The grub has four larval instars, and the larval stage occupies about 20 days. They feed on all stages of mealy bugs. Duration of first, second, third and fourth instar grubs are 3-4, 4, 4-5-7-8 days respectively. Grownup grubs are entirely covered with white wax strands. When the grub is disturbed, it exudes a yellow fluid from the dorsal surface of the body for defensive purpose.

The prepupal period is 2 to 4 days when it suspends feeding activities. The pupal period varies from 7 to 9 days. The adult spends about one day in the pupal case before it emerges. It is covered with a white powder like substance for a day. The male could be distinguished from the female by the colouration of first pair of legs. The first pair of legs in the case of male is brown and the latter two pairs being black, whereas in the female all the three pairs are black. Male to female ratio is 1 : 1. Adults are also known to attack and feed the mealy bugs. Longevity of adults ranges from 50 to 60 days and the fecundity is about 200-220 eggs.

Chapter 14

Enemies of Ladybird Beetles

Enemies of Coccinellids

Although knowledge of naturally occurring parasite, pathogens, and parasitoids of coccinellids has increased in recent years, a limited number of studies have investigated the biology and ecology of those species that attack beneficial ladybird beetles, which often contribute to the reduction of herbivore populations in managed and unmanaged ecosystem. Recent advances have been made in the biology of dipteran and hymenopteran parasitoids of coccinellids, as well as pathogenic bacteria, pathogenic fungi and nematodes. However, the biology and ecology of viruses, nematodes, mites, and fungi that attack coccinellids are largely unknown.

Anti-Predators Defence

Ladybirds display a range of general defence reactions, from escape (by flyind away, running away or dropping to the ground) to immobilization and also other more specific anti-predator adaptions. Adults of many members of coccinellidae are consipicuosly coloured, often with contrasting red and black or yellow and black patterns on their elytra. Such paterrns usually serve as aposematic (warning) colouration. Larvae and puapae may also be aposematically coloured with dark and bright areas on their surface. The function of warning colouration is to advertise ti potential predators that the bearer is unpalatable, toxic or nutritionally unprofitable. Indeed, ladybirds are often distasteful and toxic to vertebrate and vertebrate predators.

Predators of Coccinellids

Despite being distasteful and toxic, ladybid beetles have been reported to be eaten by various vertebrate and invertebrate predators. Predation by vertebartes concerns virtually all the main groups: Fish (Gomiero *et al.,* 2008), Amphibians (Cicek and Mermer 2007), reptiles (Pal *et al.,* 2007), birds (Mizer 1970) and mammals (Chapman *et al.,* 1955). Predation of Coccinellidae by birds has been analyzed in detail. The contribution of coccinellidae to the diet of birds was highlighted by mizer (1970) in Ukraine through analusis of food remnants in almost 7000 stomachs of birds belonging to 234 species. He detected ladybirds in the diet of 20 per cent of bird species.

Spiders are frequently reported as preying on ladybirds, especially web-building spiders (*e.g.* Nentwig 1983,Laing 1988, Richardson and Hanks 2009, Sloggett 2010). In contrast of web-builders, actively hunting spiders can capture their prey in more targeted manner. It was found that these predtors relatively rarely attack coleopterans because of their thick and hard cuticle. In the case of Coccinellidae, defensive alkaloids may act as additional protection.

Insects of various orders (*e.g.* Hemiptera, Diptera, Coleoptera, Neuroptera) have been reported to prey on Coccinellidae and many of them are intreguild predators. Insects not belonging to the same guild as their predatoryvladybird prey include gomphid dragonflies (Odonata: Gomphidae) (Conrad 2005), robber flies (Diptera: Asilidae) (Ghahari *et al.,* 20017) or vespid wasp (Hymenoptera: Vespidae).

Predatory insects known to huny phytophagous ladybirds include beetkes (entomophagous Coccinellidae, Carabidae, and Cantharidae), true bugs (Anthocoridae, Nabidae, Reduvidae, Pentatomidae, and Lygaeidae), neuropterans (Chrysopidae, Myrmeleontidae), butter- flies (larvae of some Noctuidae), earwigs (Forfi culidae) and ants (Formicidae) (Howard and Landis 1936, Ohgushi 1986). Ants are also known as occasional predators of predatory coccinellids (Sloggett *et al.,* 1999, Majerus *et al.,* 2007), but the majority of interactions between these two groups of insects are of a competitive nature.

Parasitoids of Coccinellids

Coccinellids are attacked in all life stages by nearly 100 species of parasitoids, which primarily belong to the orders Hymenoptera and Diptera. Some are specific for limited ranges of taxa within the Coccinellidae, while others are broad polyphages, which can parasitize members of various insect families and orders. Furthermore, there are also many secondary parasitoids or hyperparasitoids associated with the primary parasitoids.

Parasitoids of different Subfamilies of Coccinellidae

The majority of records concern four of them, Coccinellinae, Epilachninae, Chilocorinae and Scymninae, with Epilachninae being associated with the highest number of parasitoid species. Interestingly, most parasitoids reported from the Coccinellinae, Epilachninae and Chilocorinae are polyphages and hyperparasitoids, while those of Scymninae, as well as of the remaining four subfamilies (Microweiseinae, Sticholotidinae, Ortaliinae and Coccidulinae), and are mainly host subfamily and family specific taxa.

Parasitism in different Development Stages of Coccinellids

All developmental stages of ladybirds are subject to attack from parasitoids, though parasitization of eggs and adult beetles is relatively rare and the former is restricted to phytophagous ladybirds of the subfamily Epilachninae. It has been supposed that predatory ladybirds are devoid of egg parasitoids due to the cannibalistic habit of the newly hatched larvae. Successful parasitoid development in the eggs prior tobeing cannibalized is unlikely because egg parasitoids usually emerge from parasitized eggs after the non-parasitized eggs have hatched (Klausnitzer, 1969). However, newly hatched larvae of some epilachnine ladybirds are also known to cannibalize unhatched eggs in the batch (Nakamura 1976, Nakamura and Ohgushi 1981).

Parasitization of adult ladybirds, although only a few parasitoid species are involved, has received disproportionately more attention from researchers than parasitization of other stages. Such over representation is largely due to one species, the euphorine braconid, Dinocampus coccinellae, which, since the fi rst systematic studies by Ogloblin (1924) and Balduf (1926), has continuously attracted the attention of successive generations of entomologists. The predominant group of parasitoids of ladybirds is formed by the parasitoids attacking larvae and pupae. However, only very few species in this group, such as the flies *Phalacrotophora fasciata* and *P. berolinensis*, or the wasps *Oomyzus scaposus* and *Pediobius foveolatus*, have been studied in any detail. The biology and ecology of the latter species has been relatively well recognized because of its economic importance as a biological control agent of phytophagous ladybirds.

Figure 15: Parasitoid Attacking Adult Ladybird.

Hymenopteran Parasitoids of Phytophagous Coccinellids

The effects of parasitoids on populations of phytophagous Epilachninae have been documented more often than on entomophagous species. *Pediobius foveolatus* provides control of the introduced *Epilachna viginsexpunctata* (Boisduval) feeding on solanaceous crops.

Hymenopteran Parasitoids of Entomophagous Coccinellids

Parasitoid of entomophagous coccinellids is *Dinocampus coccinellae* Schrank (Hymenoptera: Braconidae). This parasitoid prefers adult coccinellids especially older females, but immature parasitoids have been found in host larvae and pupae. As solitary endoparatoids of adults, *D. coccinellae* prefers larger species of coccinellinae. Coccinellids vary in their suitability as host for *D. coccinellae*, but the mechanisms that influence realized rates of parasitism of various coccinellid species are unknown of many species.

A number of other Hymenoptera are larval or larval/pupal parasitoids of entomophagous coccinellids. *Oomyzus scaposus* (Thomson) (Hymenoptera: Eulophidae) is a widely distributed gregarious parasitoid of coccinellids. It has a wide host range that includes Coccinellinae, Chilochorinae, and Scymninae including *Harmonia axzridis*. Females lay eggs in third and fourth instars of coccinellids larvae and adult wasps emerge from pupal mummies.

Microbial Parasites

Parasites of ladybird beetles can be loosely grouped as endoparasites and ectoparasites. Roundworms (Nematoda: families Allantonematidae and Mermithidae) as endoparasites of the adult stage of entomophagous ladybird beetles. *Mermis sp.* (Mermithidae) from fourth instars of a phytophagous ladybird beetle, *Henosepilachna vigintioctopunctata* F., in India.

Parasitic Fungi

The order Laboulbeniales contains ascomycetous fungi that parasitize arthropods. Approximately 80 per cent of the 2000 described laboulbenialean species include beetles as their hosts. As far as is known, all Laboulbeniales are biotophic; that is requiring a living host for continued survival and propagation. Horizontal transmission of Laboulbeniales, likely through direct contact between adult hosts, is the common mode of perpetuation. This process is greatly facilitated when coccinellid hosts aggregate at feeding or overwintering sites. The propensity of entomophagous males to repeatedly attempt to copulation with gonad heterospecific males and females under field condtions may facilitate the spread of laboulbenialean fungi populations of susceptible species. Four laboulbenialeans species within the genus *Hesperomyces virescens* (Thaxter) has multiple hosts including *Harmonia axyridis* (Pallas). Whether *H. virescens* develop and thrive on phytophagous coccinellids is unknown.

Bacteria

Most studies on bacterial infections of coccinellids concern male killing bacteria. Male killing agents include *Ricekettsia, Spiroplasma, Wolbachia,* Flavobacteria and some yet to be identified agents. Maternally inherited bacteria kill males during

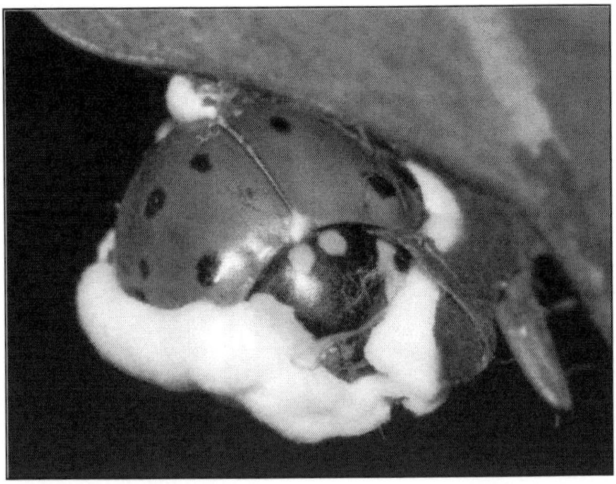

Figure 16: Adult Ladybird Infected by Fungus.

early embryogenesis, and infection provides an adaptive advantage for female offspring that feed on unhatched male eggs. Entomophagous species infected by one or more of these agents are in the genus *Adalia, Adonia, Anisosticta, Calvia, Cheilomenes, Coccinella, Harmonia, Hippodamia,* and *Propylea*.

Chapter 15

Authors Identified Coccinellids in India

Sl. No.	*Author*	*No. of Species Reported*	*Area*
1.	Subramaniam. T.V., 1923	01	South India
2.	Aiyar, T.V.R., 1942	01	South India
3.	Kapur. A.P., 1948	12	North India
4.	Puttarudriah, M. and Channabasavanna, G.P., 1953	53	Mysore State, India
5.	Usman, S. and Puttarudriah, M., 1955	48	Mysore State, India
6.	Kapur, A.P., 1972	17	Goa, India
7.	Pajni, H.R. and Singh, J., 1982	30	Chandigarh, India
8.	Pajni, H.R. and Varma, S., 1985	25	Chandigarh, India
9.	Canepari, C., 1986	36	India and Nepal
10.	Bhagat and Masoodi, 1988	04	Kashmir, India
11.	Singh, T. and Singh, V.K., 1990	04	Himachal Pradesh, India
12.	Omkar and Bind, R.B., 1993	06	Uttar Pradesh, India
13	Singh, K.C. and Singh, T.K., 1991	15	Mizoram, North Eastern State, India
14.	Singh, T. and Singh, V.K., 1990	04	Himachal Pradesh, India
15.	Omkar and Bind, R.B., 1995	07	Uttar Pradesh, India
16.	Omkar and Parvez., A., 1999	10	Lucknow region, (U.P.), India
17.	Omkar and Parvez., A., 2000	07	Lucknow region, (U.P.), India

Sl. No.	Author	No. of Species Reported	Area
18.	Sathe and Bhosale, 2001	21	Maharashtra State, India
19.	Poorani, J., 2002a	01	Karnataka, India
20.	Poorani, J., 2002b	400	Indian subregion
21.	Poorani, J., 2003	01	Karnataka, India
22.	Kandibane, M., Rahuraman, S. and Ganapathy, N., 2005	07	Madurai, Tamil Nadu, India
23.	Joshi, P.C. and Sharma, Pushpendra K., 2008	31	Haridwar region, Uttarakhand State, India

References

Agarwala, B. K. and Senchowdhuri, M. 1988. Biology and food relation of *Micraspis discolor* (F.), an aphidophagous coccinellidae in India. *J. Aphidology*. **2**(1-2): 7 - 17.

Almeida, Lucia, M., Correa, Geovan, H., Giorgi, Jose. A. and Grossi, Paschoal, C. 2011. New record of predatory ladybird beetle (Coleoptera, Coccinellidae) feeding on extra floral nectarines. *Revista Brasileira de Entomologia,* **55**(3): 447–450.

Aslam Bukero, Abdul Ghani Lanjar, Abdul Waheed Solangi, Muhammad Faisal, Shahnaz Naz Pathan and Shezad Ali Nahiyoon, 2015. Biology of *Coccinella transversalis* FAB. On *Aphis nerii* under laboratory conditions. *Sci. Int. (Lahore),* 27 (4), 3289 – 3292.

Atif, Jamila, Y., E L Husseini M.M, Agamy, E. A., and Ahmed Sayeda S., 2015. MorphologicalCharacteristics of some Species of Lady-birds (Coleoptera: Coccinellidae) in Giza Region, Egypt. *Egyptian Journal of Biological Pest Control,* **25**(2): 379-387.

Chakraborty, S. K., Bhowmik, A. R. and Biswas, S. 1996. Coleoptera: Coccinellidae, *State Fauna series 3: Fauna of West Bengal,* Part 6B: 449-493

Chakraborty, Dola, Korat, D. M. and Saneera, E. K. 2013. Correlation between coccinellids feeding on mealybugs and weather parameters. *Insect Environment,* **19**(1): 60-62.

Chakraborty, Dola, Korat, D. M. and Saneera, E. K. 2013. Relationship between predatory coccinellids and their hosts on three different pulse crops. *Insect Environment,* Vol. **19**(1):63-64.

Collier, T. and van Steenwyck R., 2004. A critical evoluation of augmentative biological control. *Bio-Control.* 31: 245-256.

Cottrell Ted E., 2017. Predation and cannibalism of ladybird beetle eggs by adult ladybird beetles http://dx.doi.org/10.1016/j.biocontrol.2005.04.008

Das, S., Roy, S. and Mukhopadhyay, A., 2010. Diversity of arthropod natural enemies in the tea plantations of North Bengal with emphasis on their association with tea pests. *Current science vol.99.no.10.*

Dixon, A. F. G., 2000. Insects predator- Prey dynamics, ladybird beetles and biological control. Cambridge University Press, Cambridge, UK.257

Frank, J. H., Kanamitsu, K. 1987. *Paederus sensulato* (Coleoptera: Staphylinidae): Natural history and medical Importance. *Journal of Medical Entomology* 24: 155-194.

Frank, J. H., McCoy, E. D., 1990. Endemics and epidemics of shibboleths and other things causing chaos. *Florida Entomologist* 73: 1 – 8

Ghosh, S. K., Chakraborty, K. and Mandal, T. 2013. Bio-ecology of predatory coccinellid beetle, Coccinella septumpunctata (Coleoptera: Coccinellidae) and its Dynamics in Rice field of Terai region of West Bengal, India. *Inernational Journal of Bio-resource and Stress Management, 4(4):71-57.*

Glorg, J. A., Vandenberg, J. N., McHugh.J.V, Forrester, J. A., Sllpinski, S. A., Miller, K. B., Shapiro, L. R. and Whiting, M. F., 2009. The evolution of food preferences in coccinellidae. *Biological control vol. 51:215-231.*

Hagen, K. S., 1974. The significance of predaceous coccinellids in biological and integrated control of insects. *Entomophaga* 7: 25 -44.

Hodek Ivo, Honek Alois. 2017. Scale insects, mealy bugs, white flies and psyllids (Hemiptera, Sternorrhyncha) as prey of ladybirds http://dx.doi.org/10.1016/j.biocontrol.2009.05.018

Kaur, Harmeet and Virk, J.S. 2011. Feeding potential of *Cryptolaemus montrouzieri* against the mealy bug *Phenacoccus solenopsis. Phytoparasitica* DOI 10.1007/s12600-011-0211-3.

Kausar Saeed, Muhammad Nasir Khan Khattak, Muhammad Fiaz Khan, Falak Naz, Adil Khan and Naveed Akhtar, 2016. Ladybird Beetle (Coccinellidae; Coleoptera) of District Buner, Khyber Pakhtunkhwa-Pakistan *J. Appl. Environ. Biol. Sci.*, 6(2S) 58-78.

Majerus, M. and Kearns, P.K. 1989. *Lady birds.* University of Cambridge. Richmond Publishing Co. Ltd. Slough, SL 23 RS, England, pp. 1-101.

Michaud, J. P., 2012. Ecology and behavior of the ladybird beetles (Chapter 11, Coccinellids in biological control). Blackwell publishing Ltd.

Omkar and Pervez, A. 2000. Biodiversity in predaceous coccinellids (Coleoptera: Coccinellidae) in India. A review. *J. Aphidology.* **14**: 41 - 66.

Perry, J., and B. Roitberg. 2005. Ladybird mothers mitigate offspring starvation risk by laying trophic eggs. *Behavioral Ecology and Sociobiology*58: 578–586.

Poorani J. 2002. An annotated checklist of the Coccinellidae (Coleoptera) (excluding Epilachnae) of the Indian subregion. *Oriental Insects,* **36**: 307-383.

Poorani J. 2004. Notes on the Coccinellidae (coleoptera) of the Indian Subcontinent, Including New Synonymies. *Journal of Biological Control,* **18** (2): 185-187.

Poorani, J. 2014. An unusual new species of *Micraspis* Chevrolat (Coleoptera: Coccinellidae) from north-eastern India. *Biodiversity Data Journal*: (2): e4112 doi: 10.3897/BDJ.2.e4112.

Poorani.J, 2017. Adult morphology. http://www.angelfire.com/bug2/j_poorani/morphology.htm.

Rahat Ullah, Faizul Haq, Habib Ahmad, Mian Inayatullah, Kausar Saeed and Shahroz Khan 2012. Morphological Characteristics of ladybird beetles collected from District Dir Lower, Pakistan. *African Journal of Biotechnology* Vol.11 (37), pp. 9149-9155.

Riddick, E.W., Cottrell, T.E., and Kidd, K. A., 2009. Natural enemies of the coccinellidae: Parasites, pathogens, and parasitoids *Biological control vol. 51:306-312.*

Saeed, Kausar, Khan Khattak, Muhammad Nasir, Khan, Fayaz, Naz, Falak, and Akhtar, Naveed (2016). Morphological Characteristics of Ladybird Beetles (Coccinellidae: Coleoptera) of District Buner, Khyber Pakhtunkhwa, Pakistan. *Pakistan Journal of Zoology,* **48**(5): 1367-1372.

Sayed, Samy M. and El-Arnaouty, S. A. 2013. Pollen grains as an additional food for Asian lady beetle, Harmonia axyridis Pallas (Coleoptera: Coccinellidae). *Life Science Journal,* **10**(3): 1755-1759.

Index